There is only one insincere animal in this world and that is man. All the others represent themselves truthfully, without any shame or pretense, expressing their feelings just as they are. No wonder we feel joy in observing these creatures. We see ourselves as we should be.

after Arthur Shopenhauer

Opossum

VIRGINA OPOSSUM

(Didelphis virginiana)

The Virginia opossum is the only marsupial found in the USs. The babies of marsupials do not fully develop in their mother's womb like the young of other animals. Instead, they are born at a very early stage and continue their development in an external pouch. (Not all marsupials have pouches. The babies of some marsupial species just hang onto their mother's nipples without the protection of a pouch.)

Opossum is correctly pronounced with the "O," although the common slang is "possum." "Possum" is not technically correct because there are other mammals from Australia that are actually called possums.

With its pointed snout, two-foot body, and foot long, hairless tail, an opossum looks a bit strange. Captain John Smith of the Pilgrims wrote of them that they have "a head like a Swine and a tail like a Rat."

Opossums are shy, solitary animals. Adults come together only to mate, and the males leave immediately after mating.

Always common in the eastern US, except in the northermost states, the opossum has spread as far north as Canada and as far west as New Mexico. It has been introduced to the Pacifc Coast. The opossum is so adaptable and has adjusted so well to humans that it has thrived while many other mammals have seen their numbers and range decline. Now it can even be found in large cities.

△ When the surviving babies are only slightly bigger than a mouse, they crawl out of the pouch and hang onto the hair on their mother's back and neck. They do not hang onto their mother's tail, as cartoons and drawings often suggest. They go back to the pouch to nurse. After another month or so they can care for themselves, and they wander off. Possums are not territorial but roam from place to place seeking food. Carrying their babies on their backs enables the females to forage without having to return to a den in order to care for their young.

THE MARSUPIAL POUCH

Virginia opossums have an external pouch in which they carry their incompletely developed young. The fertilized eggs develop into embryos in the uterus and are born 12 to 14 days after conception.

Upon birth, the baby opossums move out of the womb and crawl several inches forward into their mother's fur-lined pouch.

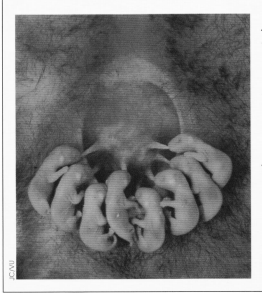

This is surely one of nature's marvels. The newborn opossums are blind, but their front legs are relatively well developed with claws which aid the trip to the pouch. They are extremely tiny, so small that 20 would fit on one teaspoon and 200 weigh only an ounce. Nevertheless, they make their way to the pouch (one trip was recorded at only 17 seconds), grab onto a teat, and hang on for two to three months. If they let go, they are doomed. There is no fleshy attachment, but the nipples swell up in the mouths of the young, thus helping them to remain attached. Sometimes more babies are born that the mother has teats (the number of teats can vary from the usual 13 up to 17), resulting in death for those unfortunate few for whom there is no nipple. Usually, in spite of the large number of teats, they are only seven to eight young in the pouch.

The young are normally found inside the pouch. The photographer in this case lifted the young outside and arranged them so it was possible to see them holding the nipples.

◁ The opossum's tail is prehensile (capable of grasping). The tail is used for balance, support, and to carry material for nest building, but usually only young opossums can actually support their weight and hang from their tail. The opossum is the only native mammal in the US that has a grasping tail, although many mammals in others parts of the world, such as monkeys, have this kind of tail.

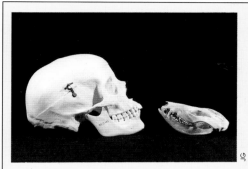

PEA BRAINS

This photo shows a human skull alongside the skull of an opossum. Opossums have the smallest brain/weight ratio of any mammal in North America. By comparison, a raccoon of the same body weight has a brain six times larger. Humans have the largest brain relative to their body size. Animals with larger brains, such as dolphins and primates, seem to be more intelligent, although the definition of intelligence is very subjective. Intelligence is usually measured by the ability to learn from humans or to solve problems devised by humans. However, the ability to adapt to their environment is the means by which mammals survive in the wild. Because of its adaptability, the opossum has not only succeeded in expanding its range, but has also managed to survive in environments that have been greatly altered by humans.

THE IMMIGRANT

Long ago, North and South America were completely separate and had different animals. Later, they became connected by a land bridge (Panama), and animals moved between the continents. Although most migration was southward, a few South American mammals were able to survive and successfully compete in North America. The opossum was one of the few that prospered. Other successful South Americans include the armadillo and porcupine.

▷ Opossum charactertics include a pink nose, naked black ears with white tips, and prominent whiskers.

△ Young opossums can be easily tamed and are fascinating to watch. However, as with other wild animals, it is best to leave them in the field.

OPOSSUM

Opossums eat many kinds of plants and animal matter, dead or alive, including garbage, so they generally have little trouble finding food. Their diet includes fruits, vegetables, nuts, seeds, eggs, worms, small fish, insects, mice, and frogs. They eat a lot of roadkill. They even eat rattlesnakes.

Although very few live to be two years of age, opossums may live seven years or more, and if not killed they continue to grow. Adults may weigh up to ten pounds, with an exceptional one reaching fifteen pounds. In captivity, one is reported to have reached a weight of more than thirty pounds.

Natural enemies include owls, bobcats, hawks, and especially dogs. Vultures often eat roadkilled opossum and there are some people that relish an opossum meal.

Male opossums often fight ferociously. As a result, many are badly scarred, have scaly tails, and are missing teeth and toes. Yet opossums constantly groom themselves and are considered very clean.

Today, the automobile probaby causes the most opossum mortality. Opossums are attracted to highways during their nightly foraging by the presence of roadkills, but they are near-sighted and have trouble seeing approaching cars.

In addition, opossums tend to freeze in the glare of headlights. This tactic serves them well in the wild because their stillness may reduce the chance of attack by a predator, but it leaves them far more vulnerable to cars. They often make no effort to get out of the way of oncoming traffic.

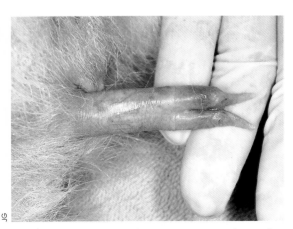

△ Female opossums have two uteri; that's the meaning of the opossum's Latin name, *Didelphis*. Male opossums have a penis with a forked tip (shown above). Scientifically, the significance, of this unusual anatomy is not clear. An interesting (but false) explanation from folklore is that opossums copulate through the female's nose.

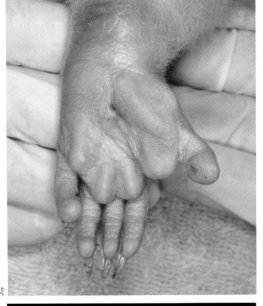

△ ▷ The shape of the opossum skull indicates the small size of this creature's brain. However, note the large number of teeth (50), more teeth than any other North American mammal except for some species of dolphins and porpoises. (Manatees have more than 50 teeth over the course of a lifetime but only 30 to 36 at any one time.)

△ This photo shows the trademark grimace of the opossum which is actually a threat display.

◁ △ The hind foot of the opossum has an opposable big toe which works like a human thumb for grasping. Note that it has no nail. With the aid of their tail, and their rear paws with opposable thumbs, opossums are good climbers. Opossum paw prints are distinctively star-shaped.

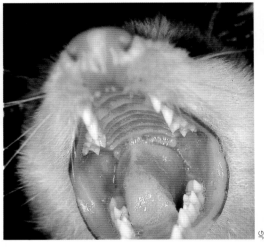

△ The color of an opossum's fur varies from light to dark but is usually some combination of gray and brown.

PLAYING DEAD

This scene happens often. A pet dog finds an opossum under the porch and drags it out. The opossum, showing its full set of teeth, is bleeding from the mouth and coated with the dog's saliva. It remains very still. The little boy that owns the dog, feeling certain the animal was dead, runs to fetch his dad, to show him this strange creature. Upon returning, the opossum is gone without a trace. The little boy earnestly declares to the father that there really had been a dead animal right there. Gently, his dad explains that "possums" play dead. A dead opossum is no fun for a dog and not very appetizing, so the animal escapes.

Feigning death is a defensive tactic used by some birds and reptiles, but much less commonly among mammals. When opossums play dead, they go into a faint, properly called "catationia," during which their heart rate and their breathing slows. They lie on their side, with their mouth open and their eyes closed. They don't respond even if they are poked sharply. This condition may last from less than a minute to several hours and seems to be a reflexive action over which the opossums have no control.

This defensive maneuver only works with predators who insist on live prey, but opossums have other defenses, too. They discourage attackers by snarling viciously and baring their teeth, and they can deliver a nasty bite. A strong odor from their anal glands also aids in their defence.

Opossums spend much time in trees, where they often enter hollows. They do not hibernate, but they put on some extra fat prior to winter which enables them to remain inactive during bad weather. However, they do emerge and their tracks can be seen in the snow, sometimes even during inclement weather. In the northern parts of their range they often suffer frostbite on their naked ears and the tip of their long, naked tail.

Opossums will fight if necessary, but often defend themselves by opening their mouth and hissing. They can swim if pressed, and one was observed swimming across a lake which was about 100 yards wide. The opossum dog-paddled on the surface, but swam under water much of the way.

A large number of ectoparasites are found on opossums. For 66 individuals studied in Indiana, these included 6 kinds of fleas, a biting louse, 15 kinds of mites, 2 kinds of chiggers, and 2 kinds of ticks. Many of these (such as the biting louse) were picked up from prey animals, but 4 of the species were host specific (confined to one type of host) and would not be in North America at all if they did not occur on opossums.

SHREWS AND MOLES

Shrews and moles are in the order Insectivora, and they do often feed on insects. However, they are not totally insectivorous, nor even the most insectivorous of mammals. This distinction belongs to the insectivorous bats. Moles spend almost all of their time underground, and shrews spend much time underground. Both moles and shrews have fine fur that can lie either way, an adaptation for an animal that spends much time running back and forth in its burrow.

Shrews and moles have tiny eyes and their ears are small and often hidden in the fur. All North American shrews have teeth tipped with chestnut colored enamel while the teeth of moles are entirely white.

Many mammals can be distinguished by their dental pattern, or dentition, and this is especially true of shrews and moles.

Unlike other small mammals, small young of insectivores are seldom seen in the field because the young do not leave the nest until nearly full grown.

Shrews

SHREWS

There are nine species of shrews in the northeastern US as defined in this book. Some of the species are quite rare, whereas the masked, smoky and short-tailed shrews may be quite common. However, unless a cat has dropped one at their doorstep, most people have never seen a shrew and are unfamiliar with them.

Most shrews spend much time underground and require soil moist enough to keep the air in their burrows 100% saturate. However, the least shrew (and a few other shrews in other areas) lives in drier situations. Most shrews feed heavily on insects and their larvae, and other small invertebrates, such as spiders and centipedes, although larger shrews often feed on earthworms and snails. Shrews constantly run about when active, poking their noses here and there. They are quite vocal, producing various chitters and clicks, and at least some of them use echolocation.

Shrews have an interesting distribution pattern, which helps them to avoid competition.

Different sized shrews often occur together but usually prey on different food bases. For example, often occurring together are the short-tailed shrew (a large shrew) which feeds on large earthworms, snails and centipedes; smoky shrews (medium sized) which feed on insect larvae, small earthworms, centipedes, and beetles; and masked shrews (small-sized) which feed on small insect larvae and spiders.

Seven of the species, those in the genus *Sorex*, have long tails, several times the length of the hind feet. The other two have very short tails, often not much longer than the hind feet. The two short-tailed shrews are easy to tell apart as they are very different in size. The northern short-tailed shrew is very large for a shrew; it is mouse-size and gray to black. The least shrew is tiny and usually brown.

The long-tailed shrews are much more difficult to distinguish from each other. Three of them (the long-tailed, smoky and water shrew) are dark gray and larger than the other *Sorex*. They are about four to six inches long. The other four are brown, less than four inches long, and very similar to one another.

LEAST SHREW (above)
Cryptotis parva

This species is a very tiny shrew, usually brownish, with a very short tail. It differs from most other shrews by living in much drier, open habitats, and in often gathering in some numbers (as many as 31) in a single nest. Most other shrews need a moist environment and are very antisocial. The least shrew usually lives in open fields, at least in the north, but sometimes occurs in marshy areas or even woods. Two least shrews in captivity were observed cooperatively excavating a burrow. One dug, and the other removed dirt and packed the tunnel walls. The least shrew is sometimes called the bee mole for it sometimes enters beehives and eats the bee larvae.

MASKED SHREW
Sorex cinereus

The masked or cinereus shrew is one of the smallest of mammals, but is widespread and is often very common. It occurs in a variety of habitats all over the northern part of North America, but like most of the other shrews, is seldom seen. A small, brown, long-tailed shrew in the eastern US is likely to be this species.

It is often found in woods, fields, brushy areas and along watercourses, but is often plentiful in wet areas where there is ample moss. Its pencil-sized burrows can be found if you look closely. Another place where shrews often run is under the lip of a bank along a road or trail. Fresh dirt and even tracks are sometimes encountered in such a situation as it provides a virtual highway for small animals.

In woods, masked shrews run about in their tiny burrows under logs or under the leaf cover, where they feed primarily on tiny insect larvae and spiders. Masked shrews have several litters each year of about two to six young in a tiny nest of grass or leaves.

PYGMY SHREW
Sorex hoyi

The pygmy shrew also looks much like the masked and other small brown shrews. It can be distinguished from the other three by the unicuspid teeth which can be seen with some magnification by pulling the lip back from the upper jaw. A cusp is a point on chewing teeth. Shrews have a very large pincher like tooth (the first incisor) in front with two lobes (the first incisor), followed by some small teeth, then the larger molariform teeth. The small teeth behind the large front incisor are the unicuspids. All the other long-tailed shrews of the northeastern states have four unicuspids of

similar size and one tiny, but visible one. The pygmy shrew has tiny third and fifth unicuspids hidden behind the second and fourth (as seen from the side), thus appearing to have only three unicuspids.

For many years this species was thought to be quite rare until pitfall traps were developed. These are sunken cans either placed into a small mammal highway (for example, under a log or lip of a bank) or placed besides a "drift fence" (hardware cloth or plastic barrier) which small mammals will often follow in the absence of a mammal highway.

SOUTHEASTERN SHREW
Sorex longirostris

The southeastern shrew is very similar to the masked shrew. Although its name, *longirostris*, means long rostrum or long snout, its snout is shorter and less pointed than that of the masked shrew. However, this character is easier to see when examining a skull than when looking at the animal (a character refers to a physical (or morphological) trait). Also, its tail is somewhat shorter and less hairy than the tail of the masked shrew. Within the area covered by this book, this species occurs only in Delaware and Maryland.

LONG-TAILED SHREW
Sorex dispar

The long-tailed or rock shrew is thought to be quite rare. It is found primarily in talus slopes (rock tumbles at the bottom of cliffs) or among moss-covered boulders. Relatively few individuals have been examined but since it is found in remote areas, it may be more common than believed. It feeds on spiders, centipedes and beetles.

7

WATER SHREW
Sorex palustris

The water shrew is a very interesting species. As its name suggests, it is found around water, and it is highly adapted for a semi-aquatic existence. It usually occurs along rocky streams, or sometimes in mossy areas around lakes. It is found throughout most of New England and farther south only in the mountains.

The other larger, dark colored, shrews have dark bellies, but that of the water shrew is silvery white. This probably makes it more difficult to observe from below when swimming. This species has a fringe of hairs on its toes which increases the surface area of the foot and traps air bubbles allowing it to walk on the surface of the water. It has water resistant, velvety fur which traps air. The animal can dive to the bottom, and can stay on the bottom as long as it keeps swimming, but when it stops, the air in the fur makes the animal pop back to the surface like a cork.

Water shrews feed on aquatic insects and other invertebrates, but will also eat various terrestrial items. They breed in spring or summer with two to three litters of about six young per year.

SMOKY SHREW
Sorex fumeus

This species occurs almost entirely in mature woods and is often quite common. It makes dime-sized burrows in overhanging lips of banks, and under rocks and logs. Like all shrews, it is very active and has a high metabolic rate. It remains active in extreme cold. The author once trapped several specimens in New York during a period when the temperature during the day never rose above 20 degrees below zero, and reached a low of 31 degrees below zero during the night.

Litters of two to eight are produced in spring and summer. This species feeds heavily on smaller earthworms, insects and their larvae, and centipedes.

TAMING OF THE SHREW

Shakespeare's famous reference to the shrew may be insulting to women and certainly would be considered politically incorrect today, but Shakespeare was biologically correct in using the shrew as an example of an extremely fierce animal.

How did a small, mouse-sized, insect-eater inspire such strong language? In old English folklore, shrews were wrongly blamed for various maladies of horses and cattle, particularly any paralysis or numbness of the legs. The following quote from a book published in the 1600s shows the unpleasant reputation the animal unfairly suffered at that time: "It beareth a cruel mind, desiring to hurt anything, neither is there a creature it loveth... they take their prey by deceit... they love the rotten flesh of a raven. If any horse... feeds in a pasture or grass in which a shrew has put forth her poison, it will presently die."

The short-tailed shrew is actually a ferocious mammal and its bite is painful. Both male and female shrews are aggressive. The word shrew was originally applied equally to both sexes of humans and also meant to berate or curse (to beshrew someone). Later, shrew came to mean only a scolding woman. Shrews in the wild produce high-pitched squeaks. Perhaps this harsh noise which could be compared to the sounds of a quarrelsome female might have inspired the modern meaning. Shakespeare enshrined forever this use of the word in one of his most popular plays without ever actually mentioning the animal.

NORTHERN SHORT-TAILED SHREW
Blarina brevicauda

The northern short-tailed shrew is the largest shrew in the US. It is about the size of a mouse and weighs about one-half an ounce to an ounce. It can immediately be distinguished from a mouse by its very tiny eyes, short, mole-like fur and short tail. If this isn't convincing, look for the chestnut-tipped teeth.

This species occurs in many habitats in the northern part of its range, but is more apt to be found in woods in more southern areas, probably because of increased moisture there.

Blarina is unique among mammals in producing poison in its salivary glands. The poison is used to paralyze prey such as earthworms and snails, its main food items. The items then will remain fresh while stored for later use. Small piles of snails, or their shells with pierce marks, can often be found around or under logs used by short-tailed shrews.

These shrews are not entirely carnivorous. They have been known to eat corn and to store caches of corn for later use. They also commonly eat tiny subterranean fungi, *Endogone*, and related forms. The fungal material looks like dirt, but the spores can easily be seen at 50 or 60 power magnification. At that power they look like little bunches of grapes, each spore with its own stem. Shrews are voracious eaters, and this species sometimes consumes up to or even more than its own weight in food per day.

The burrows of this species can be found under leaf cover or under logs. The burrows are about the size of a quarter, but are flattened from top to bottom. This differs from the round burrows of voles .

Even with our poor human noses, we can often smell Blarina from several feet away. The burrows of male short-tailed shrews are scent-marked, signaling other males in search of mates not to enter. This avoids fierce fighting which often results from confrontations between two males of this species. A submissive posture also has evolved in short-tailed shrews which results in the immediate cessation of fighting. The shrew getting beaten can lie on its back, which as in dog-fights, indicates "uncle," and ends the fight.

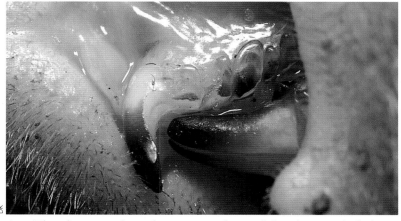

△ The sharp teeth and poisonous bite of the short-tailed shrew can paralyze its prey. The first incisor in shrews is a large, pincher-like tooth, but behind it are several upper teeth that are small and single-cusped. They are called unicuspids. Unicuspids are often used in identification and classification of shrews.

Moles are much larger than shrews and have very large, strong, outwardly projecting forelimbs (rotated with palms out) with which they dig their burrows. They can move through the soil (similar to swimming) at rates up to about one foot per minute. The burrows have main passageways which are used over and over, and side foraging passages which may be used only a few times or just once. Much more digging of burrows occurs in wet weather than in dry, and mole hills may then appear in large numbers.

Moles eyes are even smaller than those of shrews and are covered with skin. Presumably moles can distinguish light, but probably nothing more. External ears are totally lacking, but moles can hear perfectly well. Also their tactile (touch) senses are well developed. The naked snout with its few hairs is in constant touch with the parts of the burrow.

Moles eat some seeds, but most often, they hunt earthworms or grubworms (scarabaeiid beetle larvae). Their foraging habits help aerate the soil.

Moles may be active at any time during the day or night. All have one litter per year in early spring.

Unlike shrews, moles of the northeastern states are easy to tell apart. The star-nosed mole has a star of 22 tentacles on the nose and a long, hairy tail, whereas the other two have short tails and no star. The Eastern mole has a tail that is nearly naked and the hairy-tailed mole is aptly named.

△ This photo shows a hairless baby mole. Baby animals that are born without hair are usually helpless for the first period of their lives (altricial) and require a longer period of parental care. By comparison, baby jackrabbits are born fully furred and are able to run about soon after birth (precocial).

EASTERN MOLE
Scalopus aquaticus

This is the common mole of much of the southern US. In the area covered by this book, it occurs in eastern Maryland, Delaware, New Jersey, southeastern Pennsylvania and New York, most of Connecticut, and southern Massachusetts.

It is found in sandy or loamy soil. After a rain many new burrows appear as the mole moves about in search of earthworms and other items in the soil. When the weather is dry, it retreats to its deeper burrows. The nest is of leaves or grass and is usually under a stump, bush or log and has several entrances.

As in other moles, main foods are earthworms, grubs and other insects and their larvae, and other invertebrates. Ants are often eaten and this species will often enter anthills for them. Ant heads have been seen embedded in mole skins, perhaps received when ants were trying to protect their nests. Eastern moles have been examined whose stomachs are completely filled with grass seeds. This species has a litter of two to five young in spring.

MOLESKIN

The mole's fur is unique in many ways. It is among the most perfect of animal furs. The alternating round and flat segments of a single hair shaft make the coat extremely soft and capable of being bent forward and backward, thus reducing friction as the mole moves through its tunnels. Moleskin was very popular in Europe for making caps, purses, and tobacco pouches. In 1959, over one million moles were trapped for their skins in Britain alone. Since an eastern mole is only about six inches long (and European moles are similar in size), it took many skins to make anything of value. But the people who could afford these items appreciated the very fine fur. As late as the 1950s, moleskin was still popular with hikers to cover blisters. Synthetic materials have now replaced moleskin for this purpose.

HAIRY-TAILED MOLE

Parascalops breweri

The hairy-tailed mole is usually found in woods, thus its burrows may be more difficult to see as they are often covered with leaves. It can immediately be distinguished from other eastern moles by its short hairy tail. It will sometimes emerge from its burrow at night to feed. If encountered on the surface in the daytime, it appears oblivious to human presence. Like other moles, it eats earthworms, grubworms plus other insects and their larvae and other invertebrates. This mole may live four or five years. Its tunnels are extensive and may be used by later generations.

▷ This photo shows the naked snout of a mole. It is constantly wriggling as the mole uses it as a sense organ to gain information about its environment. The eyes of a mole are very poor, but its ears, although not developed outwardly, are excellent.

LIFE IN A TUNNEL

The known habits of the eastern mole revolve around its tunnel existence and its fossorial (underground) habits. A mole has huge, strong, front feet. the forelimbs are very short, which provides leverage for the powerful digging muscles. The special arrangement of bones and muscles causes the forepaws to be rotated, palms out. Its rear feet are small like those of a mouse. Using a twisting motion that may resemble swimming but which is an effective technique for excavation, the mole pushes dirt up and to the sides to form a tunnel. There is often a myriad of tunnels just below the surface of the soil. These subsurface tunnels are used as a complex system of pitfalls for trapping insects, larvae and earthworms.

The mole also digs an interconnecting system of deeper tunnels, up to eighteen inches deep. Here it leads a mostly solitary life, interacting socially once a year to mate. The dirt from the deep tunnels is pushed up to the surface and expelled in such a way that a distinctive, volcano-shaped mound is formed.

Mole burrows are often distressing to farmers, golf course greenskeepers, and home-owners, although they eat many harmful insects including grub worms. Moles sometimes damage flower bulbs and crops. Even though moles do not usually eat vegetable matter, the digging activities of the mole sometimes cause roots to be loosened and exposed to air in the tunnels. This may cause plants to wither and die.

A mole does not patrol randomly but seems to be immediately aware of events throughout its vast tunnel network. When a hole is punched into a tunnel, for example, the mole immediately goes to make repairs. Also when an insects drops through, the mole proceeds directly to the correct location to consume its prey.

A mole can bury itself in five seconds if placed on the ground. It can tunnel thirteen feet per hour near the surface. A single tunnel has been traced for 3,300 feet.

The vast network of tunnels made by a single mole may cover one-half to two acres. The tunnels are repaired as needed and may be used for up to eight years before being abandoned, probably for lack of food. The mole's tunnel network is primarily a drop-trap for insects, the main food of the mole. Prey such as scarab beetles drop into the tunnels while digging in the soil.

STAR-NOSED MOLE
Condylura cristata

The star-nosed mole has a unique star of twenty two tentacles on the end of its nose. It is the only northeastern mole with a long tail, and this species is exceedingly aquatic. It is best termed a muck inhabitant. It is an excellent swimmer and will enter water and dive to the bottom. As in other moles, its most important food is earthworms, but it eats many aquatic organisms, including fish. It is especially aquatic in winter, when its terrestrial foods are more difficult to obtain. It swims by moving its feet and tail in unison and may even swim under ice.

The tentacles are very sensitive and constantly in motion when the mole is traversing its burrows or digging in the muck. It has been hypothesized that the tentacles are electrical sensing devices to help locate food. The tentacles are kept out of the way when the animal is feeding.

Nests are of leaves or grass in a bank or hummock (a low, rounded mound) or some other raised area that keeps it out of the water. Its litter is born in April or May, and the young leave the nest at about three weeks of age. Unlike most insectivores, fairly small young of this species are occasionally seen in the field.

△ A close-up photo of the sensitive tentacles surrounding the namesake nose of the star-nosed mole.

△ This view of the star-nosed mole hunting for food reveals the long and hairy tail which is very different from the tails of the other two mole species found in the Northeast.

COMMON NAMES

Common names vary greatly from place to place and even in the same place. For example, chipmunks (Tamias) are often called ground squirrels, and ground squirrels (Spermophilus) are often called gophers. Gophers (pocket gophers, Geomys) in the deep south are often called salamanders (apparently a derivation of sandy-mounder), gopher tortoises (Gopherus) are called gophers, and salamanders are called spring lizards, or just "lizards."

A system of scientific names with rules of application has been developed (the International Code of Zoological Nomenclature, or "The Code") so that people can be sure they are discussing the same animal. There is but one valid scientific name for each species, and this name is valid around the world. Scientific information would not be very valuable if we did not know to what species it applies.

Many types of words can be used as scientific names. The most important rule is that the name must be unique. People often ask scientists who find a new species whether they are going to name it after themselves. This is simply not done. However, the name usually represents some characteristic of the animal, a geographic location, or may be in honor of another person. Scientific names don't always come out as intended, although they remain as the valid name as long as they are published in a proper scientific source using nomenclature with proper generic and specific name (binomial nomenclature) and according to The Code.

Moles are a good group to use to discuss names not coming out as intended. Two of the three moles in the northeastern US are Condylura cristata and Scalopus aquaticus. The word cristata means crested and aquaticus means inhabiting water. The star-nosed mole was named cristata because of the crested tail, an incorrect description which resulted when the tail appeared crested because the tail bone was left in the skin of the dead specimen on which the original description was based (tail bones are usually removed in specimens). The eastern mole was named aquaticus because the original collecting label indicated the specimen was collected in a cistern, however this is the most terrestrial of our moles.

Some don't like to use scientific names. That is just because of lack of familiarity with them. Most people don't have problems with the names rhinoceros, hippopotomus or rhododendron, all scientific names which are also used as common names.

△ This photo shows the extremely aquatic star-nosed mole capturing a worm underwater.

MAN IN THE NORTHEAST

During the Archaic period (8,000 to 1,200 BC), the people of the Northeast advanced from hunting deer, bison, elk and smaller game and gathering nuts, fruits, and green plants, to farming their own food. They learned how to fashion fibers into traps and nets for small animals and fish. They learned to make tools from stone, copper, bone and wood. They set controlled fires and created agricultural fields.

After the Archaic period came the Adena, Hopewell, and Mississippian periods. The most advanced period, the Mississippian, saw the construction of earthen burial and ceremonial mounds. The largest such mound in the US was located at Cahokia, Illinois and was 100 feet in height, the size of an Egyptian pyramid.

Inhabitants of the Northeast spoke one of three distinct languages: Algonkin, Iroquois, and Sioux. The Algonkins were the first to meet the English-speaking settlers, and included within their ranks were such famed persons as Pontiac, Pocahontas, and Squanto, friend of the Pilgrims. The Iroquois were famous for their strong confederation and wars with the French and English colonists. At first some Sioux lived in the Ohio Valley, but later moved to present-day Minnesota and places west and south.

It was unfortunate for the native Americans who lived in the Northeast that their area was the site of the European colonies. The native Americans fought hard to keep their land, but were defeated in the Pequot War of 1637, King Philip's War in 1675, and the Pontiac War. When the Iroquois became involved in the struggle between the English and French empires, most of them took the French side and lost power and land when the French were defeated. During the American Revolution, the Indians assisted the British, and lost again. By the end of The War of 1812, the Indians living in the Northeast had lost much of their land, and the Federal authorities sent them to live west of the Mississippi.

Today there are a few reservations in the Northeast, but the population of native Americans is small and their numbers are rapidly diminishing, although some reservations support gambling casinos, and their futures seem secure. Although the Eastern Seaboard is one of the most heavily populated places in the world, the remaining native Americans constitute a very small part.

If the US had been settled from west to east, it is unlikely that the Northeast would be so heavily populated today. The area from New Jersey to Maine is bitterly cold in winter with heavy snowfall and ice storms. The ground is frozen much of the year. Both crops and humans would have fared better on the West Coast. In spite of this, the Northeast became the first area to be heavily settled. The settlers displaced the Native Americans who had peacefully co-existed with the land. The settlers farmed and raised animals, particulary sheep for wool. This took habitat away from wild animals. The settlers hunted cougars and wolves which they regarded as only predators. Both were eliminated and the numbers of other large mammals greatly reduced.

Whale men went forth from New England to slay the mammals of the sea for their oil. Together with whalers world wide, they hunted some species into extinction. Fortunately, some whale species are making a come-back because of international restrictions on whaling. Like some whales, some of the larger land mammals of the Northeast (deer, elk moose, coyotes, otter, fisher, etc.) are also making a come-back. These increases are due to a decline in hunting and farming and to better conservation. A moose in Massachusetts is not quite as rare a sight as it might have been a few decades ago, but it is still novel enough to make newspaper headlines.

Bats

Bats are the only mammals that can fly. Flying squirrels glide, but do not really fly.

Bats belong to the order of mammals called Chiroptera which, translated from Latin, means "hand-wing." Arm-and-leg-wing might be a more appropriate name because the membrane that creates the bat's wing extends from the arm to the leg and even to the tail, although it is the wing bones that do the flying.

Bats have a short upper arm and a relatively long forearm. The thumb consists merely of a claw on the forward angle of the wing, but this claw is sharp and used by the bat to crawl and to climb up walls. It is the fingers that form the long thin bones that support the wing membranes.

The hind legs of bats are turned outward and have very sharp pointed, curved claws. These claws allow the bat to hang from the

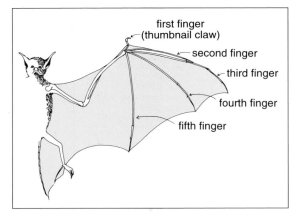

△ Part of the secret of bats' ability to fly is the modification of their hands through the lengthening of the finger bones.

first finger
(thumbnail claw)
second finger
third finger
fourth finger
fifth finger

△ This big brown bat has a complete tail membrane. Unlike the free-tailed bats, (which are not found in the northeast) its tail does not extend beyond the stretched membrane. Note that this bat has just caught a flying beetle, its major food item.

ceiling with no energy expenditure. A tiny piece of cartilage or bone, called a calcar, projects into the tail membrane from the ankle. The calcar has a keel in a few species.

The bat species of the northeast are essentially 100% insectivorous and are the main predators on nocturnal flying insects. Different bats tend to eat different types of insects. Some bats are primarily moth-feeders, some feed heavily on beetles and true bugs, and some feed on flies. However, they can vary their food greatly with changing availability or to take advantage of a favored food supply. Termites, ants, and caddisflies are particularly favored foods, and all of the bats of an area sometimes turn to them when available in large numbers. No bat specializes on them because they are not always available. It is widely thought that bats feed on mosquitoes, but they do so only infrequently. Mosquitoes are too small to make much of a meal, thus it is inefficient for bats to capture them individually. Consequently, they are seldom eaten.

Bats are long lived, have few predators, and low numbers of offspring. Most North American bats have delayed fertilization. They generally mate in the fall, but the sperms remain in the female re-

productive tract and fertilization does not occur until spring. This is often thought to be an energy saving device, allowing mating behavior in the fall when food is abundant, rather than in the spring when energy is at a premium. However, mating can take place in spring or even in winter in many species, and delayed fertilization may serve more to allow ample time to get all females mated, rather than to save energy. One or two young is the rule, but the red bat generally has three or four.

Most species of bats have been declining in recent years. The is mostly because of habitat loss because of (as with most other conservation problems) increasing human population. It is possible that pesticides and fertilizers also have played a role in declining bat populations but not enough evidence is available to clearly determine that.

Bats occasionally have to be evicted from a building, however, it should not be necessary to kill them. The only way to keep them out of buildings is to seal off their means of entry and exit. This should not be done during the summer when young are present. It is usually best done during winter, when no bats are present, or at least when they are at low population levels.

BIG BROWN BAT
Eptesicus fuscus

The big brown bat is the second largest bat of the Northeast. It is dusky brown throughout. This species has teeth large enough to give a good bite.

It is strongly adapted to man, using buildings for maternity colonies and hibernacula, and feeding heavily on crop pests. It feeds on hard insects, particularly beetles and true bugs.

Big brown bats produce two young per year (one in the west) forming maternity colonies in barns, churches, residences and other buildings or structures. Maternity colonies number up to about 600 individuals. A few males may often be found lower in the same building in which maternity colonies occur, or in nearby buildings.

Young bats often fall to the floor, at which time they make rather loud, piercing squeaks. The mother will encourage the young if it can get up onto a wall, but often will do nothing if the young remains on the floor.

Big brown bats are mostly solitary in winter, although clusters do occasionally occur. A few big brown bats hibernate in caves and mines, and a few hibernate in structures that had housed maternity colonies if the temperature inside remains above freezing all winter. Most, however, appear to hibernate in low

numbers (one to five bats) in attics in buildings not housing maternity colonies but where temperatures remain above freezing. They awaken about every two weeks during hibernation, and they often switch roosts in winter. However, they do not feed when outside in winter.

Big brown bats commonly harbor two particularly interesting parasites: a batbug similar to the bedbug, and a soft or bird tick. Mammals usually harbor hard ticks.

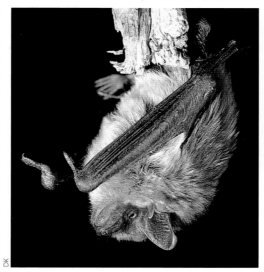

△ Bats close their mouths at rest, but during flight they must open their mouths to use echolocation and to catch food. This photo of a big brown bat shows its fierce-looking teeth which are used to capture, hold, and chew insects which often have hard-shelled bodies. Bats bite humans only in self-defense.

LITTLE BROWN MYOTIS
Myotis lucifugus

Little brown bats are quite common, but appear to be declining over major parts of their range, probably primarily because of loss of habitat.

The little brown myotis originally formed maternity colonies in hollow trees, but today they most often use buildings. They form very large colonies, sometimes up to three or four thousand individuals. The author has observed one colony containing about 6700 individuals.

Little brown bats mate in the fall, the female carries the sperms throughout the winter, with fertilization occurring shortly after emergence from hibernation. The single young is produced usually in June. Food is mostly small moths, beetles, and flies, particularly midges.

Little brown bats hibernate in caves or mines. In hibernation the bats usually form loose clusters, up to several hundred individuals. Prior to hibernation, little brown bats weigh about seven grams, and they put on about two grams of fat in fall in order to maintain themselves through the winter. They wake up on the average of once every two weeks, often fly about, then go back into hibernation. These wake-ups alone use about 75% of their stored fat, even though they are awake for a total of only about 55 hours. They may fly about inside or outside of the hibernaculum (the place they hibernate). Even if they fly outside, they do not feed, at least in the northern part of their range.

NORTHERN MYOTIS
Myotis septentrionalis

The northern myotis is common throughout the northeast. It is very similar to the little brown myotis but has longer ears and a longer, more pointed and curved tragus (the lobe at the inside of the ear). This bat is not very well known, but apparently forms maternity colonies in hollow limbs or under loose bark of trees. Northern myotis hibernate in caves. They are solitary and sometimes occur in large numbers in a cave but can not be seen because they often hibernate in tiny cracks. They feed on flies, small beetles, and moths but often eat spiders.

INDIANA MYOTIS
Myotis sodalis

The Indiana myotis is federally endangered, primarily because it hibernates in very large numbers in a very few caves. Eighty thousand individuals hibernate in one cave in Indiana, about 36,000 in one mass. Indiana bats are very similar to little brown bats. They were not even recognized as a separate species until 1928. They differ from little brown bats in having a well developed keel on the calcar, and by having very short hairs (not extending to the ends of the toes) on the tops of the toes. In contrast, little brown myotis have the keel absent or only poorly developed, and the toe hairs extend past the ends of the toes. Behaviorly the two species are quite different. In summer Indiana bats spread out and form small maternity colonies (usually less than 100 individuals) under the loose bark of trees, whereas little brown myotis usually form large maternity colonies in buildings.

LEIBS MYOTIS *Myotis Leibii*

Leib's bat is a small, uncommon myotis with black ears, a black mask, and a keeled calcar. Its habits are not well known. Females have one young per year. The few maternity colonies that have been found consisted of 12 to 20 individuals in buildings. They generally live and forage in wooded areas, and hibernate in caves. Several individuals have been found under rocks on the floor of caves and in holes underground.

RED BAT *Lasiurus borealis* (above)

The red bat is one of the most beautiful bats in the northeast and can be recognized immediately by its namesake color. The red bat is almost unique among mammals in that males and females are colored differently. The male is much richer red than the female.

Red bats are solitary and highly migratory. Red bats hang among foliage during the day. They migrate southward where they may hibernate or remain active for the winter, depending upon latitude. Sometimes flocks of migrating red bats have come to rest on boats far at sea.

They feed on a variety of insects, but especially moths, beetles, various "hoppers," ants, and some flies. They have three or four young per year, probably a reflection of greater than normal amounts of predation (for bats). Bluejays appear to prey on their young.

HOARY BAT
Lasiurus cinereus

Hoary bats are the largest bats of the northeast, having a wingspread of about 15 inches. They are light brown with the tips of the hair whitish, resembling hoarfrost. There is a white epaulet (shoulder patch) at the shoulder and the face is usually reddish. The ears are black rimmed. Females have two young. Very few males are present in the northeast in summer; males are mostly in the southwest. Hoary bats are solitary, living among the foliage. They undertake long migrations in fall mostly to central America, but some winter in the southeastern US and in Baja. They presumably mate during the southward migration. This species feeds very heavily on moths.

BAT CAVES

Many people think that bats live in caves and some do. For example, great numbers of free-tailed bats, Tadarida brasiliensis, form maternity colonies in some caves in the West, up to 12,000,000 in one cave in Texas. However, none of the bats in the Northeast form maternity colonies in caves. In the Northeast, caves are used mostly in winter for hibernation. A few males of some species of the Northeast do live in caves in summer. These males are often solitary, but congregations of these bats are called bachelor colonies. Also, bats in summer sometimes use caves as temporary night roosts where they stay for a few hours after their evening foraging and before predawn foraging.

Bats of certain species of the Noretheast do hibernate in caves: Indiana myotis, little brown myotis, northern myotis, Leib's myotis, eastern pipistrelle, and big brown bat. The most commonly seen bats in caves in the Northeast in winter are little brown myotis, followed by eastern pipistrelles, northern myotis, and big brown bats. Indiana myotis occur in tight clusters, huge numbers in a very few caves, mostly in Indiana, Kentucky, and Missouri. For example, about 87,000 Indiana myotis hibernate in Twin Domes Cave in southern Indiana, 36,000 of them in one cluster. This species hibernates in smaller numbers in some caves and mines in the Northeast.

Only certain caves are suitable for bats, because bats require fairly specific temperatures and humidities. It is important to avoid disturbing caves when bats are present. Most bats are timid and are not able to tolerate too much human intrusion. When people enter bat caves some of the bats wake up and use some of the limited quantity of fat needed to maintain them through the winter.

Grates have been used to keep people out

△ The bats shown in this photo are little brown myotis.

of caves. This has sometimes been harmful to the bats. In some cases the grates reduced air circulation, and the caves became too warm for the bats to tolerate. However, grates are now much improved and are currently the only way to protect bats from people.

Natural disasters, especially floods, have also killed large numbers of bats. Flooding can fill a cave so quickly that an entire colony of bats is wiped out.

Cave explorers have contributed to the disturbance of bats in the past. Today, however, most cavers understand and help protect bats, especially those who are members of the National Speleological Society. The protection of major hibernacula is especially important.

EVENING BAT
Nycticeius humeralis

This is a species of the southeastern US reaching north into Maryland, Delaware, and south central Pennsylvania. It is not very common in the northeastern states. It is a brown bat that looks very much like a small edition of the big brown bat. It has a forearm about the size of that of a little brown bat. It has a very short, rounded tragus.

It has a diet very similar to that of the big brown bat, feeding on beetles, true bugs and leafhoppers. It also eats quite a few moths, whereas the big brown bat generally feeds on moths only in early spring or at other times when beetles and bugs are in short supply.

It forms maternity colonies of several hundred individuals, often in buildings, but originally, and still when available, in tree hollows in bottom land. It flies south for the winter, but it isn't known where it winters, or whether it feeds or hibernates during winter. Females produce two young.

FLYING MICE

Bats are sometimes called "flying mice," but this name is very misleading. The German word for bat is fledermause which means "flying mouse" and is enshrined in the title of Mozart's opera, Die Fledermause. The French word for bat is chauves-souris, meaning "bald mouse." Although they have soft fur, bats are not closely related to mice. Bats are also much less prolific than most mice, usually giving brith only once a year to a single offspring or twins. "Flying shrew" would probably be closer to reality.

MORE BAT SPECIES THAN...

Bats make up about 20% of all the mammal species on earth. About 925 of the over 4,600 species of mammals worldwide are bats. In some tropical countries, the number of species of bats is greater than all the other species of mammals combined. In spite of their impressive presence on this planet, we know relatively little about them. Man has devoted much more energy to eradicating them than trying to understand them.

ECHOLOCATION

All bats have good vision (the expression "blind as a bat" is a myth). But since bats fly and feed at night, they rely on echolocation to track flying prey and avoid obstacles in the dark. Bats emit high-pitched sounds as they fly. Upon hearing the echoes, bats can determine the location, distance, size, shape, and texture of an object, and even its speed if the object is moving. This system is considered far more sophisticated than man-made radar.

EASTERN PIPISTRELLE

Pipistrellus subflavus

The eastern pipistrelle is the smallest bat of the northeast. It has a wingspread of only about six and one-half inches, and weighs only about five to seven grams (one-fifth ounce). It is a pretty little bat, reddish with tricolor fur (the individual hairs are dark at the base, lighter in the middle, and darker tipped).

This bat forms very small colonies, seldom more than about 30 individuals. Maternity colonies are usually in woods, often in tree hollows, but occasionally in buildings. Pipistrelles often move between two or more colonies, even moving their young. They feed on small hoppers, flies, beetles and moths. In winter they hibernate in caves and mines.

◁ During hibernation, tiny droplets of water often collect on the fur of pipistelles, giving them a silvery appearance. Water collects in a similar manner on other bats also, particularly the little brown myotis. It was thought that pipistrelles remained at rest for longer peiods during hibernation, thus allowing more water to form, but recent evidence indicates that this is not the case.

△ Bats can see, but bats in flight at night use high-frequency echolocation to locate insects and avoid obstacles. People can hear up to about 20 kiloherz. Echolocating bats use frequencies from about 20 to 140 khz. Bats can even distinguish among certain kinds of insects using echolocation.

THE BAT BOMB

The roosting habits of bats once inspired an unusual plan to bring about an early end to World War II in the Pacific. The idea was to attach very light incendiary bombs to bats and release thousands of the armed creatures over Osaka, one of Japan's major industrial cities. The bats were expected to roost in the rafters of the buildings, which during that period were made mostly of wood. Timing devices would ignite all the firebombs at once, thus creating a vast firestorm.

*The feasibility of the plan was proven one day when a few of the bats in the US escaped and burned down the control tower of the desert airbase where the project was located. Shortly thereafter, the idea was mysteriously abandoned by its Navy sponsors. There has been speculation that the project was derailed because of progress on the atomic bomb. An interesting book, **Bat Bomb**, by Jack Couffer, describes the operation in detail.*

BAT HOUSES

These simple structures are now available from conservation organizations to encourage bat breeding. Bats seem to prefer rough wooden surfaces to which they can cling easily. So far, bat houses have not been very successful in the Northeast. Many have been placed in the shade. They need to be in the sun because maternity colonies of little and big brown bats (the two species most likely to use bat houses) need heat in which to raise their young.

SILVER-HAIRED BAT

Lasionycteris noctivagans

Silver-haired bats undergo extensive migrations. They spend the summer in the northern US and Canada, and spend the winter in the southern two thirds of the US. They have often been thought of as solitary, however they appear to form small maternity colonies in hollow branches of trees.

Silver-haired bats are very dark brown, almost black, with silvery tipped hairs. The silvery coloration is sometimes heavy and obvious; sometimes it is very light. Silver-haired bats have two young per female. Little information is available, but they seem to feed heavily on true flies, caddis flies and moths.

BAT DETECTORS

Bats frequently emit chirps and squeaks that are much lower in frequency than those used for echolocation. Some of these sounds are audible to humans. Usually, these low-frequency sounds are alarm calls or sounds used for communciation between bats. The echolocation sounds are higher in frequency and most cannot be heard by humans. Scientists use receivers capable of detecting high-frequency sounds and changing them into sounds within the normal range of human hearing. Since bats echolocate at different frequencies and with different patterns, biologists can identify some bat species by their sounds.

THE BAT MAN

*Bats are probably the most maligned group of mammals on the face of the earth. Although they have been widely persecuted, they were not seriously studied until the last several decades. Fortunately, as a result of the work of Amercia's most famous "bat man," Dr. Merlin Tuttle, and other scientists and educators, many of the misconceptions surrounding bats are being corrected. Dr. Tuttle is the author of **America's Neighborhood Bats** and founder of the organization Bat Conservation International (BCI)*

BCI is the best source of informaiton about bats. This exceptional organization, headed by Dr. Tuttle, has sponsored much recent work on bats. Anybody with questions about bats is welcome to write or call (P.O. Box 162603, Austin, Texas 78716, phone 512-327-9721, fax 512-327-9724). People wanting to become members may call toll-free, 1-800-538-BATS. BCI publishes a newsletter each month to let members know what the organization is doing.

RABBITS AND HARES - GENERAL

Rabbits and hares have long ears and long hind feet adapted for jumping. Like rodents, they have a big space between their upper incisors and their molariform teeth. They also have two pairs of upper incisors, the second pair being small and directly behind the first. The fur of most rabbits is very soft.

Many people think rabbits are rodents, but they are not, although rabbits and rodents have some characters in common. Rabbits and hares have large chisel-like incisors. Also, they have vertical grooves down the front portion of the first set of incisors.

Rodents have only one pair of upper incisors, and most do not have vertical grooves on the front of the teeth. Rabbits chew from side to side, whereas rodents chew up and down.

RABBIT EARS

Cottontails have long ears, yet much shorter than those of hares. Rabbits can dilate the blood vessels in their ears to radiate heat and thus keep cool. They can also use their big ears for shade or fold them down against the body for insulation against the cold. Their ears are very tender. It is very painful for rabbits when people pick them up by the ears.

HARES VS RABBITS

The words hare and rabbit are often used interchangeably, but hares (genus Lepus) are generally larger, and have longer ears and legs than rabbits or cottontails. Hares live in more open country, and tend to outrun their predators. The young of hares are precocial (well developed at birth) and hares build no nests for them. Newborn are fully-furred, have their eyes open, and are able to run about very soon after birth.

Baby rabbits, by contrast, are altricial, meaning they are born in a rather helpless condition, hairless and blind. They are placed in a nest constructed in a depression in the ground which is lined with soft hair from the bellies of their mothers. They spend several days in the nest. The nest is covered with fur and plant material when the female is not present. The female visits the nest once or twice a day near dawn and at dusk to nurse the young. The young of lagomorphs have very little scent, which helps protect them from predators.

△ A snowshoe hair in winter colors blends with its snowy environment.

PROBLEMS WITH INTRODUCED ANIMALS

There are three species of hares in the Northeast, one native, and two introduced, and three species of cottontail rabbits, all native. Problems often occur when non-native animals are introduced to an area. An excellent case in point is the European rabbits that were introduced into the Australian region to provide a game animal for hunters. They had no natural predators and thus multiplied rapidly. Ever since the introductions, the Australian government has had to spend millions of dollars attempting to control the rabbits. The best means has been to introduce the disease Myxomatosis which kills the rabbits. However, the rabbits continually develop immunity to various strains of the disease, and it is necessary to keep developing different strains in order to control the rabbit population. Fortunately, no such problems have occurred because of the introduction of the European hare into the Northeast.

SNOWSHOE HARE

Lepus americanus

The only native hare in the Northeast is the snowshoe hare. It has large hind limbs and very large feet, hence its name. It has large ears, a short tail, and dense fur. Snowshoe hares are brown in summer, white in winter, but the tips of the ears are always rimmed with black.

The snowshoe hare is a woodland species, never straying far from woods. Its form (or hiding place) is a hollow beneath a low branch or in a tangle of vegetation. Snowshoe hares mostly sit in the form during the day, bounding away in leaps up to 12 feet in the event of danger. They can run at speeds over 30 mph.

They molt from brown to white. The molt occurs between late October and December. The spring molt occurs between March and May, at which time the coat changes back from white to brown.

Snowshoe hares thump their hind feet apparently for communication, and also emit a shrill cry when caught. Like other lagomorphs, they feed on green plant material in summer and woody vegetation in winter. Clover, brome and blue grass, dandelions and strawberry leaves are among their favored foods, but a wide variety of items are eaten.

Pre-mating behavior consists of males and females approaching each other, sniffing, jumping into the air, and urinating on each other. Snowshoe hares can produce up to four litters per year. The young are well-developed at birth and can immediately walk and even hop. After birth they split up during the day and hide, then congregate in late afternoon when the female returns to nurse them. Nursing occurs only once per day and takes only five to ten minutes.

In the far north this species has a nine to ten year cycle of population highs and lows.

The Canada lynx, a major predator of the snowshoe hare, has a nine to ten year cycle that follows that of the snowshoe hare.

Snowshoe hares have many predators in addition to the lynx, such as hawks, owls, foxes, coyotes, bobcats and the fisher. Another predator, man, has made the

△ The winter color of the snowshoe hare. Note the reddish feet.

snowshoe a favored game animal, partly because it doesn't hole up, and like the cottontail rabbit, it circles back near to the place where it was flushed.

△ Above, the snowshoe hare in summer colors, at right, in late spring.

RABBITS & HARES

EASTERN COTTONTAIL

(Sylvilagus floridanus)

The Eastern cottontail is the most common and widely distributed rabbit in the US. Most Americans are familiar with the classical literature referring to rabbits (Brer' Rabbit, the Tortoise and the Hare, and Peter Rabbit). These childhood stories no doubt help create a widely felt love affair with rabbits, cottontails in particular.

Almost everyone has seen a cottontail. The conspicuous, white, cotton-like tail is the source of its common name. When disturbed, it runs quickly to the nearest cover, but otherwise it darts about on established trails in the grass or hops about the backyard or field. It feeds mostly at night on plants and holes up during the day in thickets.

A rabbit can take a bite out of a finger if one is put into its mouth, but generally rabbits are not dangerous. They rarely bite people, with the exception of tough old males or young males that are just coming of age.

Rabbit eyes are far more light sensitive than those of man. Although rabbits do not have eyes in the back of their heads, they do have a very wide field of vision because their eyes are far back on the sides of their heads.

Female cottontails build a nest which is a depression in the ground lined with grass and fur. At dawn the mother lies down over the opening of the nest so that the young can feed.

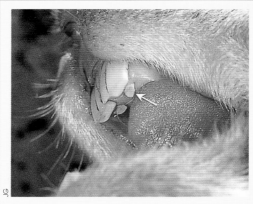

RABBIT TEETH

Rabbits have six incisors, the sharp front teeth usually used for cutting. They are arranged four above and two below. Two of the six are smaller and hidden behind the two more visible upper incisors. These two small extra incisors clearly distinguish rabbits from rodents, which only have four incisors. The two frontmost upper incisors of the rabbit are so deeply grooved that they give the appearance of four teeth. The function of the deep grooves is not known for sure, but it is suspected that they give the teeth much added strength. The function of the small pair of upper incisors is more of a puzzle, but situated where they are, directly behind the front incisors, they may add further strength to the front incisors. Also, they could be used in chewing, especially
with the lower incisors chewing against them. Rabbits feed on plant material and plant material is very difficult to chew (and to digest). Since it is difficult to chew, the incisors wear down rather rapidly. Rabbits (and also rodents) have incisors that keep growing throughout the life of the animal. This is an adaptation to keep the teeth from wearing down too rapidly. Rather than the tooth constricting into roots as the teeth mature (humans have rooted teeth), the bases of the incisors remain continually open or hollow. New tooth tissue is constantly laid down, thus explaining the continued growth throughout the life of the animal. The teeth wear down at about the same rate as they grow, about three to four inches per year, provided the rabbit has a proper diet

△ The underside of the tail is white, hence the name cottontail. Also, the white tail apparently functions like a flag which the babies can follow and also as a "flash mark," directing would-be predators to the last place where the rabbit was seen while the rabbit slinks off into the undergrowth.

▽ Baby rabbits in a nest, snuggling together for warmth

STRANGE FEEDING HABITS

Lagomorphs and some other mammals practice coprophagy (eating feces). However, they do not eat the rabbit pellets that are commonly seen deposited in piles on the ground. Rather they feed rapidly in the field, then return to the safety of their briar patch or burrow, and defecate bright-green, soft, mucous-covered pellets of essentially undigested food, and eat them directly from the anus. This adaptation helps them avoid predators and also to produce vitamins. They spend less time foraging and more time under cover. Lagomorphs are a prime target for many species of predators, thus it is advantageous for them to spend as little time foraging as possible.

MULTIPLIES LIKE A RABBIT, BREEDS LIKE A BUNNY

These common expressions refer to the fact that rabbits can have several litters per year with up to nine young per litter (usually four or five). The female often mates again shortly after giving birth. Rabbits would eat themselves out of habitat if thery weren't preyed upon so heavily. If all survived, a single pair of rabbits, together with their offspring, could produce about 350,000 rabbits in five years. However, perhaps one of five rabbits survive to reach their first birthday. Many of the larger mammals, as well as snakes, feed on cottontails.

Cottontail breeding displays are very obvious and can often be seen in a backyard. Frequently, several rabbits participate. Displays consist of face-offs, much chasing and jumping. Jumps may be staight up in the air, and the females may jump over the males.

△ This cottontail is gathering grass to build a nest. Nests are made of plant material and lined with fur pulled from the mother's pelt.

ALLEGHENY COTTONTAIL
Sylvilagus obscurus

The Allegheny and New England cottontails are very similar and until recently were thought to be one species. However, they are different enough genetically that they are now considered separate species. Both can usually be distinguished from the eastern cottontail as they have a black rather than a rufous patch between the ears, and also they lack the white patch on the forehead usually possessed by eastern cottontails. The Allegheny cottontail occurs from west of the Hudson River in eastern New York southwest through Pennsylvania and the Appalachian Mountains. The habits of the Allegheny cottontail are thought to be much like those of the New England Cottontail.

NEW ENGLAND COTTONTAIL
Sylvilagus transitionalis

The New England cottontail, like the Allegheny cottontail, has a black patch between the ears and lacks the white patch on the forehead usually possessed by the eastern cottontail. However, it is very difficult to distinguish from the Allegheny cottontail except genetically. This species occurs east of the Hudson river in New York, and Northeast through much of New England, but not in Maine. The habits of the Allegheny and New England cottontails are not well known, but are thought to be similar to those of the eastern cottontail. However, Allegheny and New England cottontails are much more secretive, seldom venturing far from cover. They make nests of fur and grasses in depressions where they have their young. Both these species are becoming increasingly rare.

THE SPREAD OF RABBITS

During the early period of European exploration, rabbits were released on uninhabited islands in remote parts of the world in hopes they would provide a source of food for sailing vessels that might visit later. Rabbits were also introduced extensively in settled areas for the purpose of sport hunting. In third world countries in times past, rabbits were often the first thing missionaries and food organizations tried to introduce, because they reproduced so quickly. In this manner, the European rabbit was introduced around the world.

Releasing rabbits on these islands was a terrible mistake. In New Zealand and Australia, rabbits nearly destroyed the sheep raising industry because they ate the grasses. Rabbits became so common in Australia that rabbit meat was exported by the ton and eaten around the world. Rabbit fur is also of commercial value, although it is not very high in quality.

ALLEGHENY VS NEW ENGLAND COTTONTAIL

The final word is not yet in on whether the Allegheny Cottontail and the New England Cottontail are the same or different species. The crux of the matter is whether they would interbreed and produce viable offspring if they occurred together under natural conditions. Although there are genetic differences between them, it is not clear whether the species criteria expressed above are fulfilled. Further research is needed on the interrelationship of these two forms.

EUROPEAN HARE *Lepus europaeu*

This is a very large hare, weighing up to 14 pounds. It has long ears, thick kinky hair, and the upper part of the tail is black. As the name suggests, this species is native to Europe. It was introduced into Dutchess County, New York in 1893, but the species may now be gone from the Northeast.

It lives in open fields, and can run at speeds to 35 mph. It feeds on herbaceous material in summer and woody material in winter. It has been known to cause severe damage to apple orchards. Young are born in January, and like other hares, the young are well furred at birth.

BLACK-TAILED JACKRABBIT
Lepus californicus

The black-tailed jackrabbit is native to the west, but has been introduced into New Jersey, on Nantucket Island in Massachusetts, and on The Delmarva Peninsula of Maryland and Delaware. It can live in prairies, pastures and cultivated fields, but on Nantucket, it often lives in rather open areas among the dunes in beach grasses.

The nest-like resting place of a hare is called a form. Black-tailed jackrabbits rest in forms during the day and come out primarily at night. Forms may be used once, or for more extended periods. Young of this species are born in slightly deeper forms which are lined with fur from the mothers breast, but the young are precocious, being well furred, with the eyes open at birth and able to move about shortly after birth.

Black-tailed jackrabbits can cover up to 20 feet at a leap. Every fourth or fifth leap is exceptional, apparently giving the animals better surveillance of their environment. They can run at speeds to 35 miles per hour for short distances. They tend to avoid water as do most rabbits, but they are good swimmers.

RODENTS

Order Rodentia

By number of species or by number of individuals, rodents are the largest mammalian order. There are approximately 2021 species of rodents. This is nearly 44% of the total number of mammal species. Rodents range in size from harvest mice weighing less than half an ounce to the beaver, which may weigh nearly 100 pounds; however most rodents are relatively small.

Rodents have only one upper and one lower pair of incisors and no canines. This leaves a wide gap, the diastema, between the incisors (front teeth) and molariform teeth (rear teeth). The incisors have enamel only on the front. The upper teeth wear against the lower ones in such a way that the softer inner surfaces wear more rapidly, keeping the tips of the incisors sharp. The incisors grow throughout the animal's life (if they didn't they would be worn away). Wear must equal growth, or the incisor may grow completely out of the animal's mouth and prevent eating, or it may curve inwards, sometimes even growing into the skull and causing death.

Like the rabbits and hares, rodents have eyes on the sides of the head, which enables them to watch for danger in front, behind, or above them.

Most rodents have four toes on the forefeet, five on the hind. Most are active throughout the year; but some hibernate in winter. A few in desert habitats even estivate (a summer sleep similar to hibernation). In both hibernation and estivation, body temperatures fall to within a degree or two of environmental temperature, and bodily functions are greatly reduced, thus conserving energy. Most hibernators burn body fat to survive during hibernation, but the chipmunks wake to eat stored food. The surface area of small animals is large in proportion to their bulk, causing them to lose body heat rapidly through radiation, thus they have high energy losses. Most are extremely active. Activity warms an animal, but also uses energy, thus rodents need large amounts of food.

Most rodents are nocturnal, although most of the squirrels are active in daytime. The abundant rodents are food for many predators. They compensate for this by having a high reproductive rate.

A few rodents are major pests of man, carrying disease, eating or spoiling stored grain and other foods, and destroying vast amounts of property. Other rodents feed heavily on weed seeds and help keep insects in check. A few rodents, such as beavers and muskrats, are exploited for their fur.

SQUIRRELS

WOODCHUCK

Marmota monax

Most people don't realize it, but the woodchuck is actually a large squirrel. Woodchucks are found in pastures, meadows, old fields, and woods. They are most active in morning and late afternoon. They are good swimmers and climbers. They will occasionally go up a tree to escape an enemy or obtain a vantage point but never travel far from the den. Green vegetation such as grasses, clover, alfalfa, and plantain form much of their diet; at times they feed heavily on corn and can cause extensive damage in a garden.

Woodchucks often give a loud sharp whistle when alarmed, then softer ones as they run for their burrow. They then often peek out from the entrance. Woodchucks make several other sounds: tooth chattering, hissing, squealing, and growling.

Woodchucks hibernate in a winter burrow after putting on a heavy layer of fat in late summer or early fall. They make a nest of grass in a hibernation chamber in the burrow. Woodchucks curl up in a ball to hibernate, and their temperature approximates that of the environment, down to about 40 degrees F from about 97 degrees. They breathe about once every six minutes, and the heart rate drops from over 100 beats per minute to about four.

The automobile, the hunter and large predators, especially the red fox, are the woodchuck's major enemies. While woodchucks can damage crop fields, gardens, and pastures, they are beneficial in moderate numbers as their feces, deposited in a special toilet chamber in the burrow, fertilizes the earth, and their digging loosens and aerates the soil, letting in moisture and organic matter while bringing up subsoil for transformation into topsoil. In New York, woodchucks are estimated to turn over 1,600,000 tons of soil each year.

Woodchuck burrows are very obvious with their large openings (eight to 12 inches diameter), and with a mound of dirt just outside the main entrance. There are often additional escape entrances, but these lack mounds. In high grass, woodchuck runs may radiate from the openings. Burrows are up to five feet deep and 30 feet long. One or more of the tunnels may terminate in an expanded chamber containing a nest. Woodchuck burrows are used by other mammals, especially rabbits and foxes, also opossums, raccoons, skunks, and even mice. Cottontails use them mostly in winter and foxes often enlarge them for a nursery den. Bones and fur around the entrance to a woodchuck burrow indicate it has been converted to a fox den.

A male seeks a mate upon emergence from hibernation. Four to five young are born in April or early May. The young open their eyes and crawl at about one month and disperse at two months.

WOODCHUCK NONSENSE

There is no truth to the famous tongue-twister "How much wood could a wood-chuck chuck if a woodchuck could chuck wood?" It is apparently just a nonsense phrase as woodchuck do not eat or throw wood.

GROUNDHOG DAY

The woodchuck is sometimes called a groundhog. This name may be a reference to the aardvark, an African animal whose name translates as "earth pig." According to legend, the woodchuck emerges from its den on February 2, Groundhog Day. If the day is sunny so that woodchuck sees his shadow, supposedly there will be six more weeks of winter. In reality, the emergence date varies greatly with latitude.

EASTERN CHIPMUNK
Tamias striatus

The eastern chipmunk is a beautiful little animal and is familiar to many people in the eastern US. Its familiar cluck or chuck note is heard over and over throughout the wood lots of the Northeast in the fall. Chipmunks are reddish brown, with two white stripes bordered by black, running down their backs. Also, they have white stripes on their faces, over and under their eyes. They are diurnal and are smaller than gray and red squirrels, and their tails are proportionally much smaller. Their smaller tails are related to the fact that they spend most of the time on the ground. Although they are excellent climbers, they don't go nearly as high as red, gray or fox squirrels. They are found primarily in wood lots, but they also occur in brushy areas, around buildings, and in parks.

They feed mostly on seeds, nuts, berries, and fruits, but they will also eat a number of other foods at times, such as tubers and bulbs, small vertebrates, insects and other invertebrates, and mushrooms. Chipmunks have large internal cheek pouches which they use to transport food. They may make many trips back and forth between a food supply and their burrow with dry foods for storage. A chipmunk with full pouches looks like it has the mumps.

Food storage is very common as winter approaches, although there is a period in late summer when the woods become quiet as few chipmunks are about. The chipmunk is a true hibernator, which means its temperature can

△ A chipmunk in its burrow

be reduced down to just above freezing. Most hibernators store fat under the skin and in the body cavity which they burn for energy in winter, but the chipmunk stores food for winter use and wakes periodically to feed. Also, chipmunks are often seen above ground in the winter, especially on warm days, and they may take food from outside if available.

Chipmunk burrows may be quite long, up to 30 feet or more, and there are several chambers for nesting and for food storage. There is no mound of dirt at the burrow openings, as the excavated dirt is scattered about. There are usually two, but occasionally up to five entrances, and one that was studied even had 30.

There are many predators on chipmunks; hawks and owls, snakes, domestic and feral cats, and weasels are some of the most prominent. Weasels and snakes can capture chipmunks in their dens, and house cats, when abundant, can reduce their numbers considerably.

GRAY SQUIRREL

Sciurus carolinensis

The gray squirrel is gray above and whitish or silvery beneath. The fur on the sides of the tail is silvery or gray rather than yellowish or orange as in the fox squirrel.

Parts of acorns or other nutshells, especially hickory, walnut, or beechnut littering the ground indicate that one of the species of squirrels has been active. In winter and spring, little holes in snow or earth where squirrels have dug up cached nuts indicate the same. Gray squirrels sometimes gnaw on maple twigs, then lap up the sap. They often use woodpecker holes as dens, especially in winter, but they make leaf nests in high tree crotches or limbs. When newly made, these are green, and hard to find as they blend in with the other green leaves. However, they are obvious in winter when they have turned brown and the leaves have fallen. Squirrel tracks resemble those of rabbits. The gray squirrel lives in hardwood or mixed forests with nut trees, especially oak-hickory forests, but also beech forest.

The gray squirrel is active all year, mainly in the morning and evening, even digging through snow in intense cold to retrieve buried nuts.

BURIED TREASURE

Researchers have various theories about how squirrels keep track of their buried nuts. It is currently believed that squirrels simply look in all the good spots, and can smell the nuts, even under snow. When snow is deep, the squirrel tunnels under it to get closer to the scent. About 85% of the nuts may be recovered. Gray squirrels scatter their buried nuts over a wide area, perhaps to prevent other creatures from uncovering a large cache.

SQUIRREL HIDE AND SEEK

Squirrels have several defenses against predators. They can freeze in an attempt to avoid notice or take cover in dens. When no hiding place is available, they race up a tree and attempt to stay on the side of the tree opposite their enemies. The squirrel in this photo has flattened its body against a tree to make itself less conspicuous.

COLOR PHASES

Although gray squirrels are usually silvery gray in color, there are localized populations of other color phases. Black and reddish phases occur in some areas, and there is an albino population in Trenton, New Jersey.

Gray squirrels feed heavily on nuts, especially hickory nuts, beechnuts, acorns, and walnuts. They transport nuts from where they find them and cache them in a new spot. Holes are dug with the forefeet, and each nut is buried individually, and tamped in with the feet and nose. Most nuts are buried near the surface, and only a few are buried more than a quarter inch deep. Buried nuts and other items are the main foods in winter and in spring, but other items such as maple flowers, tulip fruits, and maple samaras are heavily used as they ripen.

Many other items are eaten as available, including buds, bark, maple samaras, tulip tree blossoms or young fruit, corn, berries, apples and many other fruits, fungi, insects, and a variety of seeds. Squirrels constantly traverse their home ranges keeping track of potential food items. At any one time, gray squirrels usually feed on only one food, changing the item as additional food sources come along. There is a great increase of activity in fall as nuts and other items ripen, with the squirrels spending most of their active time cutting and burying nuts.

Gray squirrels make dens in trees year-round, using natural cavities, abandoned woodpecker holes, or leaf nests, especially in white oaks, beeches, elms, and red maples. The more permanent leaf nests are woven together to withstand the weather. Rough population estimates have been made by assuming 1.5 leaf nests per squirrel. Tree cavities are often old woodpecker holes or hollows where branches have broken off.

Young are born in a tree cavity den or leaf nest and females often move their young between these, perhaps to regulate the microclimate, or perhaps to escape predation or parasite infestation.

SQUIRREL TAILS

The squirrel's tail is primarily for balance in trees, but serves as a sunshade. The gray squirrels scientific name, Sciurus, comes from a Greek word meaning "shade". The tail is also an umbrella in rain, a blanket in winter, and a rudder when swimming. The tail slows descent should the squirrel fall. Tail movements are used for communicating with other squirrels.

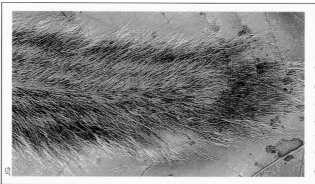

SQUIRREL FUR

The gray color of the fur of the gray squirrel is usually the result of the visual mixing of several colors of fur, including brown, gray, and yellow outer hairs, and white and gray hair underneath. In addition, individual hairs have different bands of color from the tip to the base.

SQUIRREL NESTS

Gray squirrels nest in natural cavities in dead and dying trees or make nests of leaves on the limbs of trees. These leafy nests are especially visible in the winter or early spring when the leaves are off the trees.

△ Squirrel nest in winter.

SQUIRRELS

AN EXCITING CHASE

Squirrels are commonly seen chasing one another. Young fox and gray squirrels often engage in play-wrestling with their litter-mates, leaping onto tree trunks and running along branches which builds their strength and agility. But when adult squirrels are observed chasing each other, the chase often involves three or four males pursuing a female in heat. The males often stop to fight one another. What appears to be a game is actually a serious matter of survival. The fittest male catches and mates with the female when she finally tires. This ritual insures that only the strongest males father the next generation of squirrels. It is difficult to breed squirrels in captivity because the chase is such an essential part of their mating activities. Squirrels do not mate for life, and fathers have little to do with their families. After mating they go off to pursue other females. However, the chance that the first male to mate with a female will be the father of her litter is improved because, as a consequence of their mating, the female's vagina is blocked by a waxy plug which prevents other males from impregnating her. The plug is formed when a portion of the semen coagulates due to a chemical released from the prostate gland. The plug lasts about 24 hours and then drops out, but by then fertilization has likely occurred. However, females sometimes remove these plugs. Although squirrel communities usually range over only a few acres or less, the males often travel over a wider area to find mates. This tendency helps prevent inbreeding.

SL/VU

SQUIRRELS AND BIRD FEEDERS

Gray squirrels are noted for their ability to get into bird feeders, no matter how hard people try to keep them out. There have been contests for the best way to keep squirrels out of bird feeders, and entire books have been devoted to this subject, but the perfect feeder has not yet been invented.

Even if a bird feeder is suspended from a wire, far enough from branches so that squirrels cannot leap to the feeder, they can usually make their way to the feeder on the wire itself. If the wire is covered with PVC, however, the squirrels will fall off, because the PVC will rotate when they grab it.

Flying squirrels also get into bird feeders, but their piracy takes place at night, so it usually goes unnoticed.

▷ Squirrels have many postures for feeding including sitting on their hind feet, or hanging from their hind feet, head down.

▽ Children feeding squirrels should be warned that a squirrel may inflict a painful bite if allowed to approach too close. Squirrels grab and carry food with their mouths. Although generally not vicious, they may not distinguish between the food and the hand, especially if the hand smells like food.

▷ Squirrels are exceedingly agile. They run about on branches of trees and jump from one branch to another. Hunters use the swishing sound made when a squirrel jumps between branches as a means to locate their target. Other sounds used by hunters are the sounds of nuts dropping to the ground after being cut and also the sound of teeth chewing on nuts. The large, flattened tails of tree squirrels act as balance organs and as parachutes as the squirrels scamper about in the branches. Also, squirrels have an automatic locking position of the claws, allowing them to effortlessly cling to trees.

◁ The hind legs of squirrels can rotate in their sockets to allow acrobatic postures such as the one shown in the photo at left and also the hanging feeding position shown at the top left corner of this page.

FOX SQUIRREL

Sciurus niger

The fox squirrel is the largest of the tree squirrels. Fox squirrels occur in western New York, western and southern Pennsylvania and south, but not in New England. They like oak-hickory forests, but often live in more open or park-like forests than gray squirrels. One or two tree holes per acre are needed for good habitat.

The fox squirrel is most active near dawn and dusk. Fox squirrels use tree holes and leaf nests, with each squirrel usually using three to six active nests. Leaf nests are used more when fewer tree holes are available. Leaf nests can be up to a foot in diameter. They are lined with shredded material, and have a side entrance.

The fox squirrel eats many hickory nuts, plus other nuts, acorns and seeds. They also eat the fruit of tulip poplar, the winged seeds of maple trees, corn which is ripening along wooded areas, buds, berries, and some fungi. Fox squirrels sometimes feed heavily on underground fungi, especially in the southeast. The fungi are located by smell and have large numbers of spores which are dispersed by the squirrel through defecation over several days and over a wide area. Some of the fungi contain nitrogen-fixing bacteria and are beneficial to germination and growth of trees. There is thus a close relationship between the forest, the fungi, the bacteria and small mammals. The fact that this relationship is destroyed by clear-cutting constitutes a powerful but little understood argument against this practice.

The fox squirrel spends much time in trees feeding or cutting nuts or sunbathing on a limb or in a crotch. In summer it uses a leaf nest in a tree crotch; in winter it lives in a nest in a tree hole, often with several other squirrels. Most nuts are buried individually and some in two's and three's. Some are stored in caches in tree cavities. More than one squirrel may store nuts in the same general area, sometimes even in the same tree hole, where home ranges overlap.

Winter mating "chases" are begun by males, who are ready to copulate before the females come into heat. A "chase" consists of several males following rather than actually chasing a female often throughout the day, with mating occurring with one or more of the males. The two to four young are born in late February or early March, but second litters are sometimes produced into September. The young remain in the nest for nearly two months, and first venture to the ground when nearly three months old. The eyes open at about one month.

△   Fox squirrel in the snow.

△   Squirrels sometimes lie flat on a limb to rest and to become less visible to predators.

RED SQUIRREL

Tamiasciurus hudsonicus

Other than the flying squirrels, this is the smallest tree squirrel in the Northeast, and also the noisiest. It is usually ignored by hunters because of its small size. Piles of cone remnants and small holes in the ground indicate this species has been at work. Red squirrels may be abundant in any kind of forest or around buildings.

The red squirrel is active all year, although it may remain inactive for a few days in inclement weather. Some people think that red squirrels drive off and even castrate gray squirrels. They do chase them sometimes, apparently because of the territorial nature of red squirrels, but they do not drive them off, and they certainly do not castrate them.

In conifer forests, this species feeds heavily on pine seeds and it leaves piles of cone remnants. In the fall, it cuts green pine cones and buries them in damp earth, sometimes up to a bushel per cache. Red squirrels are like other North American tree squirrels in that they store food in caches in the ground or in hollow trees.

Like all tree squirrels, red squirrels are opportunistic feeders, moving through their home ranges and trying many different foods. Food items include hickory nuts, acorns, beechnuts, and other nuts; fruits of tulip, sycamore, maple, elm; berries; bird's eggs; young birds; fungi, even amanita mushrooms which are deadly to man. Red squirrels harvest maple sugar. They bite into maples, let the sap ooze out and the water evaporate. Fresh sap is only 2% sugar, but when the squirrels later visit the dried sap, the sugar content is about 55%.

The red squirrel makes a nest, often of shredded grape bark, in a hollow tree, hole in the ground, or crotch of tree branches (like the leaf nests of gray squirrels). Females are in heat for only one day in late winter. At that time a female will allow males on her territory, and after animated nuptial chases, mating occurs.

Vocalizations include a slightly descending, drawn-out, rather non-musical trill which can be heard for some distance, and a chatter of various notes and chucks.

△ Note the distinctive black tail band and the white eye-ring. The summer coat (shown here) is separated from the underparts with a black line. Ear tufts are present.

△ A red squirrel shreds a pine tree cone.

△ The winter coat of the red squirrel.

△ Baby red squirrels, only one month old, in an apple tree.

NORTHERN FLYING SQUIRREL
Glaucomys sabrinus

Flying squirrels are small brown squirrels with very soft fur, colored rich brown above and white below. There is a loose fold of skin between the fore and hind legs, called a patagium. Flying squirrels have large black eyes. The two species of flying squirrels are difficult to tell apart, but they can usually be determined by the southern flying squirrel being smaller and grayer, with the bases of the belly hair usually gray. Bases of the belly hairs of the northern flying squirrel are usually white.

Probably the best way to locate flying squirrels is to tap on dead stubs containing woodpecker holes. If present, the squirrels will peak out, and if further disturbed will emerge and run up the tree, where they often will flatten against the trunk, usually on the side away from the intruder. Finally, they may run up to the end of one of the tallest branches, before gliding to another tree.

Northern flying squirrels are usually found in coniferous forests, mixed forests, and sometimes in hardwoods where old or dead trees have numerous woodpecker-type nesting holes, especially in stumps six to 20 feet high with holes near the top. The northern flying squirrel is found in the Northeast in New England, New York, and south through the Appalachian Mountains.

Flying squirrels are quite common, but because they are nocturnal, they are seldom seen except by a woodsman who cuts the tree in which they live. The flying squirrel makes a nest of shredded bark in tree hollows, or sometimes a bird house, or sometimes it uses a leaf nest in summer. Also, it may cap an abandoned bird's nest to provide a temporary shelter.

Flying squirrels spread their legs and stretch their flight skin in gliding from tree to tree, pulling upright at the last instant to land gently. While southern flying squirrels sometimes enter torpor (a state of reduced body metabolism), northern flying squirrels do not. Northern flying squirrels are active throughout the year and forage on the ground a great deal.

They feed primarily on lichens and subterranean fungi, such as *Endogone* and its relatives. The trees, the fungi and the mammals have a relationship in which all participants benefit; the mammals defecate, dispersing fungal spores and nitrogen fixing bacteria which help trees obtain nutrients and water. Clearcutting should be avoided, because it breaks up this association necessary for germination and proper growth of the forest.

The northern flying squirrel also eats various nuts and seeds, as well as insects, and probably stores much food for winter use.

△ A northern flying squirrel in its den.

Flying squirrels mate in late winter. After a gestation of about 40 days, two to five young are born in spring, often in a hollow stump or limb, sometimes in a bark nest in a conifer crotch. There is apparently only one litter per year. Newborn weigh only 1/8 to 1/4 ounce.

This species spends much time foraging on the ground. The southern flying squirrel is more aggressive and sometimes displaces the northern. The southern selects hardwoods, while the northern selects conifers, when the two occur together. The main predators are owls, but many other predators such as foxes and cats capture flying squirrels at times.

◁ A southern flying squirrel. Note the loose folds of skin which are stretched for their gliding flights.

SOUTHERN FLYING SQUIRREL
Glaucomys volans

Flying squirrels are the smallest tree squirrels, the only nocturnal ones, and the most carnivorous of the northeastern squirrels. This is the smaller of the two flying squirrels and usually lives in deciduous forests. It inhabits the entire eastern US except for northern New England and the southern tip of Florida. It has a silky coat, grayish-brown above and white below, with the hairs usually white to the base. The gray-brown tail is flattened from side to side. As in the northern flying squirrel, there is a loose fold of skin, the patagium, between front and hind legs. Flying squirrels have large black eyes. The two species of flying squirrels are very similar and hard to tell apart, but the northern species is generally larger and richer brown, and its belly fur is usually gray at the base.

Woodpecker holes are favored nest sites, but the southern flying squirrel may build a summer nest of leaves, twigs, and bark. This nest is similar to that of gray or fox squirrels, but it is only about eight inches in diameter. There may be primary nests, plus many secondary nests used for temporary shelter. Flying squirrels know their home range very well, and can hide in a hollow tree or under loose bark, or dart into another hiding place wherever they happen to be. They may also use old nests of birds or other squirrels.

Typical dens are dead tree stubs eight to 20 feet high and containing woodpecker holes. The nest cavity is often lined with shredded bark or other plant fibers. Some retreats are used exclusively for defecation, and buildups of humus can, over time, concentrate to half a meter. Active all year, flying squirrels may remain in their nests in extremely cold weather. In winter, several may den together in one tree hole.

When flying squirrels feed on hickory nuts, they leave the nuts with a smooth opening at the thin end. White-footed mice make two or three openings, red squirrels make a ragged opening, and fox and gray squirrels crush the nut.

Tracks are similar to those of the red squirrel, but slightly smaller. In snow, it is nearly impossible to distinguish tracks of the two.

FLYING SQUIRRELS CANNOT REALLY FLY

Flying squirrels cannot really fly. Their "flight" is a gentle and graceful downward glide. When they spread their feet, a large flap of skin stretches between their front and rear legs on each side. Spread out, this skin looks like a parachute with a tiny body at the center. After launching themselves from the top of a tree, flying squirrels usually land on the trunk or branch of another tree 20 to 60 feet away. They then climb to the top, so they can jump and glide again to the next tree.

Flying squirrels leap from a head-down position high on the trunk of a tree, or from a horizontal limb, and then extend their legs to stretch their membrane taut. They usually glide up to about 60 feet. A "flight" of about 100 yards was once observed down an especially steep hill.

A flying squirrel holds its long tail over its back when it runs, but when it glides its tail is extended straight back and is used as a stabilizer and rudder. Flying squirrels have good control over the path of their "flights." They can make sharp turns, bank to avoid obstacles, and even execute downward spirals. The flight membrane is made of sheets of muscle. These muscles can be tensed or relaxed, so the glide can be controlled by varying the tension of the muscles as well

△ The arrows point to a strip of cartiledge which extends from the wrist. It supports and widens the flight membrane. Also note that compared to the gray squirrel which has a fluffy tail, the flying squirrel has a flatter tail. The flatness of the tail makes it more useful as a rudder during glides.

as by moving the tail. When preparing to land, a flying squirrel swings its tail forward, causing its body to turn upward, as this *reduces speed and allows the squirrel to land face up against a tree trunk as shown below.*

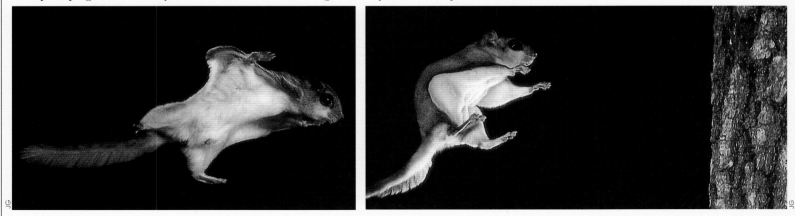

JUMPING MICE

Jumping mice are known for their ability to jump long distances. Their long tails help them balance in mid-air. Great leaps may help them to escape immediate danger, but their main tactic is to remain motionless and rely on camouflage. The woodland jumping mouse is an especially strong jumper and although it usually walks from place to place, it can leap up to eight feet when startled or in a hurry.

MEADOW JUMPING MOUSE
Zapus hudsonius

There are two species of very long-tailed (meaning that the tail is much longer than the head and body) and large-footed mice in the northeastern US: the meadow jumping mouse and the woodland jumping mouse. Both have deeply grooved upper incisors. The grooves can be seen with the naked eye. The meadow jumping mouse lacks a white tail tip, it is more yellow in color than the woodland jumping mouse, and it has a small tooth at the front of the upper molariform tooth row. The woodland jumping mouse has a white tail tip, is redder in color, and lacks the small tooth at the front of the tooth row.

The meadow jumping mouse occurs in meadows, marshes and sometimes in woods. It favors grassy areas with good ground cover. It is often found at the edge of ponds although it does not require wet ground in order to survive. The two species do not commonly overlap. The woodland jumping mouse almost never occurs in open areas, but the meadow jumping mouse fairly often enters woods. The two are most apt to be found together in patches of *Impatiens* (touch-me-not) along streams just inside of woods.

Meadow jumping mice emerge in late April or early May from their grassy nest in a hibernaculum usually at the end of a tunnel in a bank. The males emerge on the average about two weeks before the females and mating occurs almost immediately. Most females have two litters of two to nine young, and births occur in three peaks: June, July, and August.

Food is of many plant and animal items, with insects (caterpillars, beetles and many others) constituting about half the food in early summer, and a progression of various seeds becoming prominent later as they ripen. The seeds of Timothy grass are often obtained by cutting off the base of the stem, reaching high to pull it down, then cutting it off again, until the head is reached. This behavior results in a crisscrossing stack of match-length pieces of stem, with grass flower tops heaped on top. At other times the mice will climb the stem, cut off the head, and bring it to the ground to eat. About 12% of the diet of this species is of the subterranean fungus, *Endogone* (see next account).

By October 20 or so all meadow jumping mice have disappeared. The adults will have put on about six grams of fat to maintain themselves through the winter. About one third of the mice will survive

△ A meadow jumping mouse. Note brownish stripe above and yellowish sides. The woodland jumping mouse is redder.

the winter, while many will perish before spring. This is especially true of the young from late litters that did not have time to put on an adequate layer of fat.

The meadow jumping mouse can make jumps up to about three feet. However, it is more apt to hide in clumps of grass, or to run along the ground rather than jump.

CLASSIFICATION OF JUMPING MICE

Jumping mice are not very closely related to most other rodents, but have often been classed (primarily on skeletal characteristics) somewhere between squirrels and true mice. However, they have some physical characteristics in common with the jerboas of African and Asian deserts, and are currently placed in the family Dipodidae. The jerboas are long-tailed, large-footed jumping animals. A common behavior is that they enter deep sleep during adverse periods. However, jumping mice live mostly in northern woods or meadows, not in deserts like jerboas, and they hibernate in winter, whereas jerboas often estivate rather than hibernate. (Estivation is very similar to hibernation, but occurs during hot or dry rather than cold periods.)

WOODLAND JUMPING MOUSE
Napaeozapus insignis

The woodland jumping mouse is one of the prettiest small mammal species. It is reddish in color and the last inch or so if its tail is white. It does make surprisingly long jumps, up to six or eight feet and sometimes even farther. It lives almost entirely in woods, and like its counterpart, the meadow jumping mouse, feeds on a variety of plant and animal foods.

Two foods are particularly interesting, the fungus, *Endogone*, and seeds of touch-me-not, *Impatiens*. The fungus *Endogone* forms clusters of tiny spores in the soil. Under the microscope at low power they resemble dirt, but at 40 or 50 power they look like little clusters of grapes, each spore with its own stem. Fungi are not supposed to contain much in the way of nutrients, but *Endogone* and related fungi make up about 35% of the diet of the woodland jumping mouse. Seeds of *Impatiens* are heavily eaten when available. In ripe seeds the endosperm is bright turquoise blue. The entire stomach of animals (meadow or woodland jumping mice, white-footed mice) that have eaten this item may be blue. The seeds can be eaten by humans. They taste like walnuts.

This species emerges from hibernation in May and usually has only one litter of two to seven young per year. Woodland jumping mice put on about seven grams of fat in the two weeks prior to entering hibernation. This species may drum its tail on the ground when frightened.

MICE AS PREY
Many small mammals are active mostly at night and for this reason are seldom seen and often overlooked, yet they play an important role as an abundant source of food for meat-eating mammals, birds, and snakes. The great horned owl (in the photo above) preys almost exclusively on small rodents such as the deermouse it is holding in its talons.

NATIVE RATS & MICE

The rodent family, Muridae, is the largest, most successful, and most adaptable family of mammals of the world. This family contains the most species (1326), the most individuals, and occupies the greatest area of any mammalian family.

Scientifically, there is no significant difference between rats and mice other than size. There are three groups (subfamilies) of rats and mice in northeastern US: Old World rats and mice (Murinae); New World (native) rats and mice (Sigmodontinae); and the voles and lemmings (Arvicolinae).

There are five species of native rats and mice in the Northeast: two rats (rice and wood rat) and three mice (eastern harvest mouse, deer mouse, and white-footed mouse). Since they are native to the western hemisphere they are described as New World mammals. All have long tails and prominent ears and eyes, and all but the rice rat have white under-parts. They also have the cusps of the molariform teeth in two rows from front to back.

RICE RAT *Oryzomys palustris*

The rice rat is a southeastern species but is included here because it reaches northward into southeast Pennsylvania and New Jersey. Note the grayish belly which is unlike that of the other native rats and mice of the Northeast, which have white bellies.

As its name suggests, this is a species of marshy areas. It lives in a variety of wetland habitats, including salt and fresh water marshes, and it is common around flooded fields. It is a source of food for owls, hawks, foxes, skunks, weasels, mink, and water snakes.

This omnivorous rat is nocturnal; it only feeds at night. It eats seeds and fruits primarily, but also insects, and even small crustaceans and fish. Where rice is grown, it does indeed feed on the grains.

It breeds throughout the year in the south, but probably not in New Jersey. The nest is a grapefruit-sized ball of grass-like plants above the waterline in the cattails or bulrushes.

This species produces many runways and also feeding platforms of vegetation bent over the water. Feeding platforms are often strewn with crab remnants. The species is an excellent swimmer above or below the water and feeds primarily on aquatic plants, but it also feeds on crabs and other invertebrates, insects, fruits, and the subterranean fungus, *Endogone*.

EASTERN HARVEST MOUSE

Reithrodontomys humulis

Like the rice rat, this is a southeastern species, but reaches north into Pennsylvania and Delaware. It is the smallest of the Northeast mice, weighing at most only a little over ten grams. It has white under-parts, bordered by pink buff. It can be distinguished from all other long-tailed mice (except jumping mice) by the deeply grooved upper incisors. Jumping mice are much larger, have much longer tails and also much larger hind feet.

Eastern harvest mice live in old and brushy fields. They produce litters of three to five young from spring to fall after a 21-22 day gestation. Young become independent and are weaned in about three weeks.

Harvest mice live mostly on seeds but insects and other invertebrates are also eaten. Excess seeds are cached. Nests of grasses are built in low herbaceous or woody vegetation.

ALLEGHENY WOODRAT

Neotoma magister

The Allegheny woodrat looks much like a very large white-footed mouse, with its white belly, long tail and large ears. It occurs in much of Pennsylvania and in northwestern Connecticut south to Maryland and Delaware.

It is becoming increasingly rare in the northern parts of its range, at least partly due to a brain roundworm of the raccoon, *Baylisascaris*. Eggs of the worm exit in the feces from the raccoon. The worm is not harmful to the raccoon, but may be lethal to many other species, including woodrats (and humans).

The woodrat is apparently extirpated (gone entirely from one area, but present elsewhere, in contrast to extinct, which means the entire species is gone everywhere) in New York, and its status is not clear in Pennsylvania.

It lives in caves and crevices on bluffs and cliffs. One can find fecal pellets, stick houses, middens, debris piles, and cuttings. Middens contain stored food only, and are used in winter. They contain items such as red cedar or tree-of-heaven fronds or leaves, maple samaras, and pokeberry fruit.

Debris piles help to close off the openings to the den. Houses can be in a cave, a crevice, or sometimes in an old building. The house may be closed over at the top or open like a bird's nest.

Up to three or four litters, of two to six young (usually two) may be produced per year. Food is mostly green vegetation, but Allegheny woodrats will also eat fruits, nuts, seeds, and fungi. This species makes few sounds except foot-drumming and tooth-chattering.

UNLIMITED POPULATION GROWTH OF HUMANS

Many chapters in this book mention the population cycles of various animals. What are the consequences for humans of continued population increase? Space (and the resources included within the space) is always limited. Populations are healthiest when they are considerably below the maximum number the area can support. Overcrowding among laboratory rats leads to increased stress, fighting, killing of subordinates and cannibalism. In the wild, when animals exceed the carrying capacity (the number of individuals that can be supported in an area without causing undue ecological harm), major problems occur, such as starvation, ecological degeneration, disease, and infighting. The result is a massive and traumatic population decline. This knowledge gained from animals should pro-

vide a lesson to humans because humans show many of the same characteristics as other mammals when subject to overcrowded conditions.

The signs mentioned above all indicate that humans have surpassed their carrying capacity within the world, yet the present world population of humans is doubling every 40 years or so. Implications from other animals are clear. Humans are on a path that can only lead to disaster for us and especially for our descendants. We have the intelligence and the technology to avoid disaster. Unfortunately many people refuse to seriously consider the problem, and worse, some suggest solving the problem by producing more food and finding jobs to support more individuals. This ap-

proach can buy some time in the short term, but avoids the ultimate problem. Since the size of the earth is limited, and since the present population of 5.7 billion people is already over the carrying capacity and doubling every 40 years, this approach will only bring additional grief in the long term. The only sensible approach is to limit and eventually reduce the human population, and to base the world's economies on a stable rather than ever increasing number of people. Increasing human population is the root of most of the major problems on earth, and solving this and the related ecological problems should be a top priority, otherwise the world will be very different fairly early in the next century.

DEERMOUSE

Peromyscus maniculatus

There are two forms of the deermouse: the small short-tailed field form and the long-tailed woodland form (shown at left). Characteristics of the white-footed mouse (such as size, hind foot length, and tail length) are quite similar to those of deermouse making it hard to tell one from the other. Making things even more complicated are young individuals. The young of all three types are slate gray above and white below.

A great variety of plant and animal foods are eaten, and items are cached, although deermice do not cache nearly as much as white-footed mice. Two newly recognized diseases are carried by this species: Lyme, or deer tick disease, and the hanta virus. Immature deer ticks are often found on deer and white-footed mice, and hanta virus spores may occur in deermouse or white-footed mouse feces.

Breeding is from spring to fall. Gestation is about 21 days. About three to four young are produced by the prairie deermouse, about three to seven by the woodland forms.

WHITE-FOOTED MOUSE
Peromyscus leucopus

The white-footed mouse is one of the most common small mammals in the Northeast. It lives mainly in woods, but is also found in brushy areas and in fields, especially near woods. It is commonly found in buildings in the Northeast, especially in more wooded settings; whereas the housemouse is often found in barns and in buildings in cities and towns.

This species is a pretty reddish brown mouse, with large ears and bulging black eyes. It has white under-parts and a tail somewhat less than half the total length. In southern Pennsylvania and in Delaware this species could be confused with the eastern harvest mouse, but that species is very uncommon and has deeply grooved incisors. The housemouse is grayish underneath.

The white-footed mouse is most apt to be confused with, and sometimes is very difficult to distinguish from, the deer mouse, *Peromyscus maniculatus*. The deermouse has two forms in the east: the prairie deer mouse, a small mouse of sparsely vegetated fields of western New York and Pennsylvania; and the woodland forms of the deer mouse which occur in much of the woods of the Northeast. The Prairie deermouse has small feet (usually 18mm or less) and a short tail, usually much shorter than half the total length of the animal. The woodland deer mouse is somewhat larger than *P. leucopus* and has a tail more than half the total length.

The white-footed mouse feeds upon a variety of seeds, nuts, berries, insects (including many catepillars), and other invertebrates. It commonly eats seeds of *Impatiens*, which turns the stomach bright turquoise (see woodland jumping mouse account), pits of black cherries, and hickory nuts. Cherry seeds, hickory nuts, acorns, or many other items may be stored in caches. The caches are often in a hollow limb under natural conditions, but may be in a boot, under the hood of a car, under a pile of blankets, or in some other hidden place. Pits are removed from the cherry seeds by biting one round hole in them. Two or three round holes are chewed into a hickory nut to get to the meat.

The white-footed mouse breeds from spring to fall and sometimes even into the winter. Several litters of three to seven young are born after a gestation of 21-24 days. The juveniles are slate gray. White-footed mice spend much time in trees. Nests may be in the ground under a tangle of roots, in a bank, under or in a log, in a covered over birds nest, in a dresser drawer, or in many other protected places.

White-footed mice are mainly nocturnal. Agitated animals will drum their feet. This species does not routinely hibernate, but individuals sometimes do.

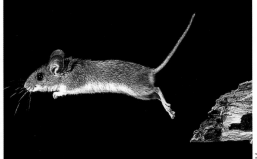

VOLES

Arvicolinae

Voles and lemmings are mostly small, short-tailed animals with small eyes and inconspicuous ears. Most feed primarily on green vegetation, which is very hard on teeth. They have adapted to this condition by having rootless teeth. The bases of rootless teeth do not close into roots when they mature. Rather, they remain open and continue to lay down tissue at the base, which means that the teeth continue to grow throughout the life of the animal. Voles have loops and triangles on the faces of the molars, a loop in the front and a loop in back and triangles between. The number and shapes of the loops and triangles are useful in species identification. Many of the voles chew off sections of grasses or grass-like plants, leaving the stem pieces in a pile, often with the flower parts on top. Voles live in quite an array of habitats, woodland, meadow, tundra, and even aquatic.

MEADOW VOLE

Microtus pennsylvanicus

The meadow vole is one of the most common small mammals of the Northeast (along with white-footed mice and short-tailed shrews), and is often the most common mammal in grassy areas. It has a relatively long tail for a vole, much longer than that of the other Northeast voles with the exception of the rock vole.

The meadow vole feeds almost entirely on green vegetation and rootstocks of various grasses and herbs. Grasses (Timothy, bluegrass, panic grass and others), yarrow (*Achillea*), plantain (*Plantago*), and dandelion (*Taraxacum*), are among its favored food plants.

The best way to find populations of this species is to look for the numerous runways and frequent piles of cuttings on the ground among the grasses. The cuttings are inch to inch-and-a-half long pieces of stem, usually of grass, crisscrossing one another. These are very obvious when numerous meadow voles are active, but you need to get on your hands and knees and part the grass to see them. They are apparently made as the vole cuts off a grass stem, grasps it above the cut, pulls it down, then cuts it again, with the process repeated until the head is reached. However, they don't feed on mature seeds, only the tender young developing seeds. Meadow voles apparently eat almost their own weight daily of green vegetation or rootstock.

Besides surface runways, this species constructs an underground burrow system. Spherical nests of shredded grass are produced. They may be found in burrows or in grass clumps or under objects such as logs or hay bales in summer, and often on the surface in winter when snow cover is present.

The meadow vole is extremely prolific. It produces several litters per year of up to 11 young per litter from early spring until late fall (throughout the year in the South). In one case 13 litters were produced in a single year by one female. Assuming litters of six (averaging three females each), then one female could yield a total of 4, 16, 64, 256, 1024, 4100, 16400, 55,600, 222,400, and 889,600 breeding females and a like number of males after 1, 2, 3 .... to 10 litters if all females survived and each produced 3 female

△   A meadow vole eating the bark of an apple tree in Maine.

offspring. Thus a single female could yield about 1.8 million mice in a single season if all survived. Luckily, meadow voles are eaten by a host of predators including domestic cats, foxes, coyotes, hawks, owls, snakes and others.

The meadow vole stamps its hind feet like a rabbit when alarmed. It also has a series of vocalizations used for threats. There seems to be a three- to four-year population cycle in this species, although some authorities question this.

THE MEADOW VOLE NEST

These photos show the grassy nest of the meadow vole. When meadow voles are abundant, these nests are easy to find. Look where grass cuttings are numerous or under objects like boards or logs on the ground.

POPULATION CYCLES

Population cycles have been suggested for a number of species but are often based on limited data. These cycles have been fairly well demonstrated for only a few species, but they include the meadow vole, the Canada lynx, the snowshoe hare, and among birds, the ruffed grouse. In the far north, where the Canada lynx feeds primarily on the snowshoe hare, the Canada lynx cycle follows the snowshoe hare cycle.

Some disputes about cycles concern the definition of the word cycle. How regular must highs and lows be to call them cycles? Population highs and lows clearly occur every three to four years in meadow voles, and every nine to ten years in ruffed grouse and in northern populations of snowshoe hares and lynx. The use of the term cycle cannot be defended if one has to assume that the highs and lows are always of equal magnitude. They are not. The highs some years are much higher than in others, and the lows also vary greatly. However, population highs and lows clearly occur and they are on a rather regular basis in these species.

All species have highs and lows, but not on a regular, predictable basis. Since this population behavior has been termed cyclic and it is different than in most species, it needs a name. There should be no problem in using the term wildlife cycle, while being aware that the regularity is not the same as that which might be expected by physicists or mathematicians.

The cause of cycles is also the source of much debate. Biologists, without success, have tried to link wildlife cycles with cycles of the moon, sunspots, rainfall or vegetation cycles.

Psychological stress factors are one likely cause of cycles, although this still doesn't indicate why these particular species are cyclic rather than just having increases and decreases (irruptions and catastrophes), such as are found in many other species. Following this idea, the greater the number of individuals in an area, the more encounters, and thus the greater the stress that will occur. This stress then causes hormonally induced factors leading to decreases in populations such as reduced number and size of litters and increased age at first breeding. During the increase phase of the cycle, population characteristics are in place that lead to greater numbers of individuals. There are more litters, more young per litter, and females breed at an earlier age.

RED-BACKED VOLE Clethrionomys gapperi

This vole, as its name suggests, is usually quite reddish. It lives in the woods, especially around rotting logs and mossy boulders, and also in swamps and bogs. The woodland vole also occurs in woods and is slightly reddish, but it has much softer mole-like fur and a much shorter tail. The woodland vole spends most of its time in burrows, whereas the red-backed vole spends much time on the surface. Rock voles and bog lemmings also occur in the woods, but have grizzled brown fur. Also, the bog lemmings have faint grooves on the face of their upper incisors, and the rock vole has a yellowish nose.

The red-backed vole has litters of two to eight young from early spring to late fall. It is a woodland species, and feeds on vegetation, such as false lily-of the-valley and goldthread. Pieces of plant material can usually be found scattered here and there along its runways. Subterranean fungi also form an important part of its diet, especially in fall. Red-backed voles often store bulbs, tubers, nuts and parts of stems in their burrows for later use. They often hop, but also can run. This species occurs throughout much of the Northeast.

RUNWAYS

Runways are paths through the environment which serve like highways for small creatures to travel to feeding areas. They may go through thick grass, under logs, or under banks. They are kept worn down by continuous use.

ANOTHER SPECIES OF VOLE?

There is a population of voles on Muskeget Island, Massachusetts, which some biologists have recognized as a separate species, and have named Microtus breweri. Its habits and behavior are similar to those of the meadow vole. However, this author does not believe there is sufficient evidence to consider them a separate species.

WOODLAND VOLE
Microtus pinetorum

The woodland vole has a very short tail, not much longer than the hind foot, as in the bog lemming. The bog lemming, however, has much more grizzled fur, and also, lightly grooved incisors. The woodland vole spends more time underground than most other voles, and accordingly has very short, soft, mole-like fur which can lie either way. This allows the animal to more easily go backwards or forwards in the burrow.

Large amounts of food are stored in underground caches in the burrow system. Burrows are constructed by digging with the forefeet and incisors, and then pushing back the dirt with the hind feet. This species eats mostly green vegetation and roots and tubers, supplemented by subterranean fungi, fruit, and a few insects.

Woodland voles appear to live in small colonies or communities, but the entire area may be suddenly abandoned, for no apparent reason.

Especially in the lower Hudson River Valley of New York, woodland voles can be a problem in orchards as they girdle the young trees, although much of the damage presumably by woodland voles is probably by meadow voles. In orchards, they can be quite easily controlled by cultivation, since vole burrows are so close to the surface.

The woodland vole was previously known as the pine vole, a very poor name as it seldom lives in pine areas, although the specific scientific name also refers to pines.

ROCK VOLE
Microtus chrotorrhinus

The rock vole, originally called the yellow-nosed vole, because its nose is often yellow, is not very common. It lives around boulders and in talus slopes in the mountains. The rock vole feeds on bunchberry, grasses, mosses, blueberries, some caterpillars, and subterranean fungi. Remnants of leafy plants can be found scattered in its home range (the area it occupies in its day to day travels). It deposits its dark gray-green fecal pellets in specific latrine areas.

BOG LEMMINGS

There are two bog lemmings in the Northeast. Both have grizzled fur and look like meadow voles with exceedingly short tails. Both have lightly-grooved incisors Neither species is often seen. The two species are quite similar, but the northern bog lemming has rust-colored patches at the bases of the ears which the southern bog lemming lacks. Both species of bog lemmings produce bright green fecal pellets, and leave piles of match-stick-length grass-cuttings. They feed on sedges, grasses and other herbaceous plants.

SOUTHERN BOG LEMMING
Synaptomys cooperi

This is the more common bog lemming that occurs throughout the Northeast, but is rarely seen, even by biologists. Also, it very rarely occurs in bogs. It is much more apt to be found in woodland burrows, or in grassy fields.

NORTHERN BOG LEMMING
Synaptomys borealis

The northern bog lemming (shown above) is apparently very rare and localized, and in the Northeast, occurs only in Maine and New Hampshire. This species occurs in bogs, spruce woods, alpine meadows and tundra.

MUSKRAT

Ondatra zibethicus

This rather large aquatic rodent has a long, scaly tail which is flattened from side to side. The fur is glossy brown, very thick and fine. The hind legs are enlarged and the feet are webbed. Muskrats live in almost any situation with abundant aquatic vegetation, such as ponds, marshes, lakes, streams and rivers. They feed on aquatic plants, especially cattails, sedges, rushes, water lilies and pond weeds. They sometimes eat freshwater clams, crayfish, frogs and fish. Muskrats may be seen at any time of the day or night, but they are especially active between dusk and dawn.

Muskrats build lodges or houses of aquatic plants, most often of cattails. These lodges may be up to about eight feet in diameter and five feet high. They usually have one chamber and house one adult muskrat, although several may live in one house for a time except during the breeding season. If the water is too deep or conditions otherwise are not right for lodges, muskrats dig a burrow into a bank. Bank burrows may have several chambers and several entrances, some of the entrances opening under water. Muskrats often have feeding platforms consisting of cut plant material in the water or on the ice. There are usually floating pieces of plant material present where muskrats are active.

Much of a muskrat's time is spent in water. It propels itself with its large webbed hind feet and steers with feet and tail. It can swim backward, and can stay underwater for long periods. One individual stayed underwater

for 17 minutes, came up for three seconds, then submerged again and stayed under for another ten minutes. Muskrats can chew or feed underwater, since the lips close behind the incisors.

Breeding occurs from late spring through early fall. The young are born naked, but are furred and can swim and dive at about two weeks of age. They are weaned and driven away by the mother when they are about two months old.

Raccoons, mink and humans are the most common enemies of muskrats. Muskrats can damage ponds and levees through their burrowing activities.

Three species of Old World rats and mice have been introduced into North America from Europe: the black rat, the Norway rat, and the housemouse. They have long, naked tails, dark-colored bellies, and teeth with cusps in three longitudinal rows.

BLACK RAT
Rattus rattus

This rat entered North America in 1609 with the colonists at Jamestown. It was more common earlier, and is much less known than is the Norway rat. It was here before the Norway rat arrived, but the Norway rat is more aggressive and often displaced the black rat. Its tail is very long, longer than half its total length, while the Norway rat has a tail less than half its total length. This species is common on ships, in the south, and along the coast to the north.

Black rats are excellent climbers, and they often live in the upper stories of buildings. Litters consist of two to eight young, after a gestation of 21 to 26 days. Like the Norway rat, this species carries a number of diseases including plague, or black death.

△ Damage from the black rat shows why this rodent is also called the roof rat.

◁ The tail of this black rat is longer than its body. The tail of the Norway rat is shorter than its body. This is the best way to distinguish the two species.

▽ The tail of this black rat (and also the Norway rat) appear hairless at first glance but actually have sparse, short hairs.

BLACK RAT VS NORWAY RAT

Although the black rat is far more common in the tropics, the Norway rat has proved to be more adaptable in cooler places, especially in cities. As the Norway rat spread through North America and Europe, it drove the black rat from its favored habitat on the ground and the lower levels of buildings. Thus, the black rat has become rare or has disappeared, especially in northern areas.

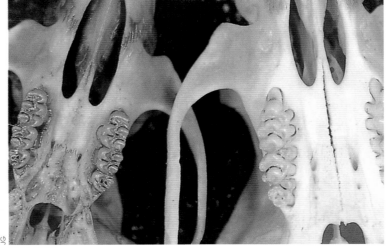

◁ These skulls show how different the tooth patterns of rodents can be. These differences are very valuable for identification. The teeth of the black rat have three cusps while those of the native rice rat have only two. In both cases the cusps aid in crushing seeds and other food.

HOUSEMOUSE
Mus musculus

Housemice are typically assumed to live in houses, but the great majority of them actually live in cultivated fields. There they are essentially beneficial as they feed upon weed seeds, insects, and insect larvae. The housemouse usually builds its nest in a burrow, but is loosely colonial and shares its range with other individuals, sharing escape holes, feeding and toilet areas.

With its high biotic potential this species is able to invade a habitat and rapidly build to rather large numbers. Numbers then remain stable until the field is harvested or plowed. If the population becomes too large, females often become infertile.

Housemice are almost nomadic as they move between fields as they are plowed or harvested. Occasionally, populations get way out of control, such as in the mouse plague of 1926-27 when populations reached the incredible number of about 82,000 per acre in the Central Valley of California.

The housemouse is native to the Old World and arrived in North America with the early setters, first in Florida with the Spanish, then in the north with the French and English.

△ The house mouse has a long tail and seems to prosper in man-made structures. A mouse found in a human dwelling may be a house mouse (if the house is in a wooded area, it is likely to be a white-footed mouse). The house mouse is not as serious a pest as the black rat and the Norway rat, but it can contribute to the spread of such diseases as food poisoning, murine typhus, and the plague. It consumes some stored human food and contaminates far more. It also destroys woodwork, furniture, upholstery, and clothing.

THE BAD RAT

The harm which rats do to man is unmatched in the mammal world. Rats are thought to eat about one-fifth of the world's crops each year. Norway and black rats consume vast quantities of food and grains and contaminate much more than they eat. They also chew on electrical wires, sometimes starting fires. On a worldwide basis, the direct damage caused each year by these two species amounts to billions of dollars. Rats also spread some 40 different diseases. They carry the black death or bubonic plague which killed about one-quarter of the population in medieval Europe. Bubonic plague is still a threat in many parts of the world. It is estimated that rat-borne diseases have taken more lives in the past 1,000 years than all the wars ever fought. Furthermore, rats frequently bite people. Over 10,000 rat bites are recorded every year in the U.S. alone. Yet rats are hard to control. Not only are they highly adaptable, but they are among the most prolific mammals on earth. A single pair, under ideal conditions, can have up to 15,000 descendants in one year.

THE GOOD RAT

On the plus side, rats contribute more to medical research than any other mammal. The common laboratory rat used all over the world is a white mutant version of the Norway rat. Nearly all new drugs are tested on these lab animals.

△ The white laboratory mouse above is an albino version of the house mouse. At left is a breeding facility for laboratory mice.

◁ Rats are often admired for their survivorship. They are among the most numerous and successful mammals on earth.

CONTROLLING RATS

Poison only works for the period it is used. As soon as the poison is no longer supplied, the population will spring back. Cats are an ancient deterrent, but cats can only kill a limited number of rats. Although the average rat lives no more than a year, rat populations can quickly multiply to match the size of the available food source. Rats can be permanently eradicated only by eliminating their habitat. This means eliminating garbage and carelessly stored food supplies, and the environmental conditions that provide suitable food and shelter for the animals.

NORWAY RAT

Rattus norvegicus

The Norway rat is also called the common rat, brown rat, water rat or sewer rat. It has caused more problems for man than probably any other wild mammal. It eats or ruins huge amounts of stored foodstuffs such as grains, has caused floods by burrowing through dams, and has caused fires by gnawing on matches. In addition, rats are an important carrier of diseases such as typhus, spotted fever, bubonic plague and tularemia. Rats played a prominent role in plagues which killed about a quarter of the people in Europe in the early 14th century.

The Norway rat is especially known for feeding on stored food. It can puncture tin cans to get at food in the home, and it eats soap, meat, vegetables, poultry, and especially grain. Much more food is contaminated with the rats' fecal matter and urine than what the rats themselves consume. Such food cannot be cleansed and must be destroyed.

The Norway rat apparently arrived in North America about 1776 in boxes of grain brought by the Hessians hired by Britain to help fight the American colonists.

The Pied Piper in the German nursery rhyme rid the town of Hamelin of rats by charming them into the Weser River where they drowned. This fable probably originated as a result of observing rats in migration. Overcrowding can lead to mass migrations, such as occurred in 1727 when huge numbers

crossed the Volga River in Russia. Many of them perished, but many survived.

Norway rats construct a series of interconnecting underground burrows. The tunnels are two to three inches in diameter, up to 1.5 feet deep and up to six feet long. Contained in the system are several entrances, some concealed exits, and chambers for nesting and feeding. Burrowing is accomplished with the front incisors. The lips close behind the front teeth

△ The Norway rat shown here is a laboratory rat.

(as in all rodents), which prevents dirt from going into the mouth and down the throat. Dirt is pushed under the animal, then backwards with the hind feet.

Food consists of essentially all human foods, grain, garbage, insects, and other items. Norway rats enter water to obtain aquatic plants and climb trees to get fruit. They kill chickens and eat their eggs.

△ The Norway rat is a good swimmer and climber. It has a repetoire of calls—squeaks, whistles and chirps. It has a gestation of 21 to 26 days, may mate right after giving birth, and it breeds throughout the year. Also, the young can breed when about four weeks old. About five litters of seven to eleven young are produced, but a female can produce up to 12 litters in a year, and up to 22 young per litter, thus this species has a huge biotic potential (ability to produce numerous offspring).

THE GOOD NAME OF RATS

The activities of the few pestilent species of the genus Rattus have adversely affected the reputation of all mammals with the word rat in their common names. However, most of the economic and health problems caused by rats are associated with only two species.

The black rat is found throughout the world and is responsible for the spread of disease and great agricultural losses. The Norway rat, known over most of the world for its destruction of property and stored food, is a threat to the safety of people, domesticated animals, and wildlife. Under certain conditions several other species of rats have become pests in local areas.

BUBONIC PLAGUE

Plague is a worldwide problem, not just limited to Europe. Historically, outbreaks have been divided into several major time periods. We are currently in the fourth outbreak, which began in Yunan, China in 1860. In 1665, plague killed about half the people of London and during one 21 year period, it killed about 9,000,000 people in India. In the early part of this century, serious outbreaks in the US occurred in San Francisco, Galveston, and New Orleans. There are currently less than 100 plague victims every year in the US, mostly in the southwestern states. The invention of streptomycin in 1948 greatly reduced the death rate among plague victims.

The disease reaches man through bites from infected fleas carried by rats, and not directly from the rats. The fleas carry bacteria which causes the plague (also called the black death). When a plague-infected rat dies, the fleas carrying the bacteria attempt to move to another animal or to man.

The effect of plague in Algeria during the Second World War is described in the classic book by Albert Camus, The Plague. The following excerpts give some idea why plague caused so much fear and panic.

"He saw a big rat coming toward him... trying to get its balance...(it) spun around on itself...and fell on its side. Its mouth was open and blood was spurting from it...There were more and more dead vermin in the streets, and the collectors had bigger truckloads every morning. On the fourth day the rats began to come out and die in batches...The old man...started feeling pains in all sorts of places--in his neck, armpits, and groin...at noon the sick man's temperature had shot up to 104, he was in constant delirium and had started vomiting again. He kept repeating, "Them rats! Them damned rats!" On that day 40 deaths were reported... The residences of sick people were disinfected; persons living in the same house were to undergo quarantine; burials were supervised by the authorities...The telegram read: Proclaim a state of plague stop close the town."

△ Note the short tail (shorter than the body) which distinguishes the Norway rat from the black rat.

OLD WORLD (EUROPEAN) RATS IN AMERICA

The rats and mice which were inadvertently brought to the Americas from Europe are the black rat, the Norway rat, and the house mouse. These are the rats and mice that people most often encounter because of their tendency to live in human dwellings. The generally harmless native rats and mice are much less frequently seen because they usually live in natural habitats. So when people think of rats and mice, unfortunately, it is usually the harmful exotic species.

CLEAN RATS?

Rats are considered dirty, but they spend a good part of each day grooming themselves. After eating in some filthy place such as a garbage can, they clean themselves. Notice that the Norway rat above has taken its whiskers into its mouth for cleaning. Rats spread disease through their droppings, which wind up in grain and in human food. Thus, rats are very unsanitary, but they are not unclean. In fact, if a human touches a mouse or rat, the rodent will immediately proceed to clean itself vigorously.

AMERICAN BEAVER
Castor canadensis

The beaver is the largest North American rodent. It weighs up to 70 pounds, and sometimes even more. It is highly aquatic. The hind feet are webbed. Valves cover the ears and noses of beavers when underwater thus keeping water out of these openings, and the nictitating membrane (a thin, clear membrane) protects the eyes while underwater.

Much of the western US was explored in order to trap beavers, but beavers were nearly extirpated in the eastern US by the turn of the century. However, they were reintroduced in many places and given protection. Now, they are thriving once again, to such an extent that they are often a nuisance.

The beaver is known for its engineering skills. It builds dams across streams to raise the water level, lodges in which to live, canals in which to move larger logs, and trails over which to traverse land. The loud slap of a tail on the water is often the first sign of the presence of beavers.

The beaver lodge is built on the bottom of the pond or on an island or peninsula. It is constructed of branches and the trunks of saplings, held together with mud. There is a large room inside with a large platform, but no nesting material. There are two or three exits to the outside. The platform is used for feeding and for sleeping.

Dams are built of logs and sticks, and secured with mud, sod and stones when available. Additional branches and more mud and sod are laid on top of these until the desired height is reached. The water behind the dam must be deep enough so that it will not freeze to the bottom. Branches are stored for food by anchoring them in the mud at the bottom of the pond. These will be used in the winter, but additional branches will be cut in winter when the beavers can get out through

the ice.

When beavers live along a river or any place where the water is too deep they will make a den by burrowing into the bank (they are sometimes referred to as bank beavers) and then they often pile some branches over the entrance in the bank. Beavers are primarily nocturnal, but much lodge and dam repair occurs in fall and at this time they may be much more active during the daytime.

A beaver can cut a six-inch tree in 20 minutes. The cutting may be done by one or two beavers. However, contrary to popular belief, they do not have any control over which way the tree falls. Tree cutting is for two purposes, to get logs and branches for construction, and to get food. The main food of beavers is the inner bark of wood. Beavers eat many kinds of wood, but quaking aspen is the most sought. Aspens are usually removed before other species are used. However, a large number of other species may be eaten: willow, maple, alder, apple, birch, and many others. Conifers are seldom eaten. Many herbaceous aquatic plants are eaten in summer including pondweeds, pond lilies, algae, burr reed, duckweed, and many others. Also, terrestrial plants are

sometimes eaten, including alfalfa, clover, and a variety of other species.

Beavers have extra digestive glands and a three-parted cecum (appendix) containing microflora which helps to digest vegetative material. Coprophagy occurs, allowing beavers to finish their meal in the safety and convenience of their lodge. However, as in other herbaceous feeders, beavers must eat a great deal in order to obtain enough nutrients.

Beavers form tight family groups including the parents and the previous litter (and sometimes a few older animals), and all work together. Most colonies contain five or six beavers with a maximum

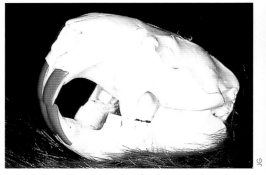

△ This beaver skull shows the large incisor teeth which grow throughout the animal's lifetime.

△ Beavers have split nails in the second toe of the hind feet which they use as combs. This is unusual, as most mammals groom with their teeth and tongue.

◁ A beaver lodge.

of about twelve. Territorial boundaries are marked with mud and vegetation to which is added urine, castorium (a strong smelling, oily substance from special "castor" glands), and the anal gland contents.

Beavers are among the few mammals that are monogamous. However, males sometimes mate with more than one female. Mating occurs between January and March in the Northeast and there is one litter per year. At birth, the young are well-developed, well-furred and their eyes are open. They remain in the lodge for a month, then venture out to swim and feed. The young are weaned in about six weeks, but can take longer.

Because of their size, beavers have few enemies other than man and his traps. There is less trapping now because fur prices have declined along with the demand for fur. The decline of trapping removed one of the few controls on beaver population growth.

Beavers have an interesting assemblage of ectoparasites (parasites of the fur), all of them host specific (found only on beavers). There is a small beetle, which trappers refer to as the beaver flea. In addition, beavers have an assemblage of tiny mites, all of which have clasping legs which help them hold to individual hairs. There are at least 15 different species of these mites on North American beavers. They are classified in four groups, based on sucker plates of the males, with members of the various groups occurring on different parts of the beaver. The many species have probably evolved (and are presently evolving!) as a result of the mites moving onto various parts of the beaver and the beavers moving about.

The beaver manipulates its environment probably more than any other species except man. Besides creating a wetlands and lodge for itself, it creates habitat for fish and other aquatic mammals, and also for aquatic birds, amphibians, reptiles, invertebrates and plants. Like most other mammal species, there is still plenty to be learned about them.

△ The broad beaver tail is used for steering in water or for slapping the water to signal alarm. It is also used as a prop to help beavers stand upright on land when they are gnawing on a tree. It may have a role in fat storage and body heat regulation.

CUTTING TREES

Beavers use trees for both food and shelter. The inner bark of trees is their main food, but beavers will eat the succulent parts of many plants, avoiding the dry bark and the wood. To construct their lodges and dams, beavers cut down trees by chewing through the trunks. A beaver can fell a six-inch diameter tree in about 20 minutes.

Contrary to popular belief, beavers have no control over which way a tree falls. The tree simply falls toward the side they chewed the most. Most trees are probably chewed more on the downhill side, since that's the side the beaver encounters first when coming from the water, so it might appear that the beaver intends to have the tree fall toward the water.

FELT COWBOY HATS

Felt cowboy hats (made by Stetson and others) are stamped with a series of X's on the inside, indicating how many beaver pelts it took to make the hat. Felt is hair compacted under enormous pressure. The finer the hair, the more it can be compressed, and the higher the compression, the higher the quality of the felt and of the hat. The average cowboy hat ranges from one to four X's, meaning that hairs from one to four pelts were used to make the hat. 4-X hats can cost as much as $500. A rare 20-X hat could be worth $3,000. The felt hats that stimulated the beaver trade in the early days were elegant hats, not cowboy hats. Nowadays, most felt hats are made from compressed wool; very few are made from the fur of beavers.

SCENT GLANDS

The beaver is one of the very few mammals with a cloaca, an anatomical structure common among birds, reptiles, and amphibians. The cloaca is a chamber through which both the urine and the fecal material empty. In the case of the beaver, the scent glands also empty through the cloaca. Both male and female beavers have two sets of scent glands: anal glands and castor glands. Both of these glands are located near the tail, under the skin where they are not visible except as a bulge in very large individuals. These glands are similar to the anal glands of dogs and cats and the scent glands of skunks. The beavers' scent glands produce a musk which they spread on trees and "scent posts," piles of sticks, mud, and grass about 12 inches in diameter, in order to alert other beaver colonies to their territorial claims. The odor can be noticed by humans at close range but is not nearly as strong as that of a skunk.

PORCUPINE
Erethizon dorsatum

There is only one species of North American porcupine. It is a large bulky animal with a high arching back, short legs, and stiff quills on the rump and tail. There are long guard hairs in the front half of the body. The feet have unique soles with small, pebbly-textured fleshy knobs and long curved claws. Porcupines have five kinds of hair: under-fur, guard hairs, vibrissae (whiskers), bristles under the tail (useful for climbing), and quills. Quills can be up to 4 1/2 inches long, and an eighth inch in diameter.

In the Northeast, porcupines are found in New England, New York, and most of Pennsylvania. Porcupines live in the woods. They strip large patches of bark from tree trunks and limbs leaving numerous tooth marks. There are accumulations of droppings inside entrances to crevice shelters and at the bases of single trees where porcupines have fed for a long time.

Favored food trees can be recognized by their stunted upper branches, bare bark, and by the accumulation of nip-twigs below. Nip-twigs are the terminal branches of trees which have been cut off and dropped to the ground after the leaves or buds have been eaten, leaving only short stubs from the main stem. Often, the trees near porcupine dens are low and stunted due to continual pruning.

On the ground, the porcupine has a slow waddling walk, which along with its obvious dark and light pattern, signals its identification to would-be predators, most of whom avoid it. It prefers to retreat or ascend a tree rather than confront an enemy. However, on its body there are about 30,000 quills. The quills are modified thick hardened hairs, solid at the tip and base, and hollow for most of their length. They have tiny backward projecting barbs at the free end. They are loosely attached to a sheet of voluntary muscles beneath the skin. The porcupine cannot throw its quills, but can erect them. When necessary, it lowers its head, and lashes out with its tail. If the tail strikes, the loosely rooted quills detach and are driven into the enemy, whose body heat causes the barblets to expand and become ever more firmly embedded.

Although the porcupine is usually able to defend itself, it is still to its advantage not to fight, and an elaborate warning system has evolved which tends to avoid strife. A porcupine always attempts to keep its backside to potential adversaries, and on this side there are obvious dark and light markings which serve as a first warning. Second, porcupines chatter their teeth once or numerous times if necessary. Third, when under extreme stress, the porcupine produces a pungent odor. It can cause the eyes to water and nose to run when

in a confined area such as a porcupine den. If all warnings fail, the porcupine can erect its spines and use them defensively.

A mouthful of spines is very painful at the least, and depending on where they enter, quills sometimes blind the victim, prevent it from eating, and can even cause a predator to starve. Offensively the porcupine can flick its tail, causing short quills to enter the flesh of the attacker. Porcupine quills can be very dangerous, as they are greased which helps them to continue inward. They sometimes even can cause death.

Porcupine tracks are distinctive. The toe points in, almost like that of a badger. The pebbled knobs on the soles leave a stippled impression and there are long claw

marks far ahead of the main prints. The foreprints, including the claw marks, are about two-and-one-half inches long, and the hindprints are well over three inches long. The trail may sometimes be blurred when the belly and tail brush the ground as the animal walks.

The porcupine is solitary and is active year-round. In very cold weather, it may den up in a rocky crevice, sometimes with more than one animal in a den. It is mainly nocturnal, and may hole up during the day in a hollow tree or log, in an underground burrow, or in a treetop. It is an excellent climber, although slow and deliberate.

The fisher is the one animal that has adapted to killing porcupines, but even a fisher occasionally receives a fatal injury. The fisher attacks a porcupine only on the ground, by circling it and running in and biting at its face. The porcupine keeps turning and will try to keep its face towards a tree or other object. Sometimes the fisher will climb the tree then come back down, forcing the porcupine away from the tree. The porcupine is finally killed when it weakens after repeated attacks to its face. Although it is a difficult kill, a porcupine is worth it, because it will keep the fisher in food for several days.

The porcupine is one of relatively few rodents that is a strict vegetarian. Its main food is of woody vegetation, but it will eat leaves, twigs, and green plants (such as skunk cabbage, violets, grasses and clover) and apples. In winter, it chews through the rough outer bark of various trees, including spruce, pines, fir, elm, basswood, beech, sugar maple, and cedar to get at the inner bark (cambium) on which it then mainly subsists. Hemlock is also eaten, but only the needles and small branches since the bark is rich in tannins. Red maple is not eaten because of its high tannin content. It also feeds on conifer needles in winter. At least in the Catskill Mountains, porcupines gain weight in summer and lose it in winter. There is a whole progression of foods through the seasons, differing between localities and even between porcupines.

Like bats and snakes, the porcupine is often hated, though often wrongly so. It may kill trees by stripping away the bark. Also, it is well known that the porcupine desires salt. Its gnawing for salt may damage buildings, furniture, tires and tool handles that have absorbed human perspiration.

The porcupine mates in October and November, a time when it is especially vocal, giving a variety of squeaks, groans, and grunts. One answer to the question "How do Porcupines mate?" is "Carefully." Actually, mating occurs in the same fashion as in other mammals, but not until the female is sufficiently aroused that she relaxes her quills before raising her tail over her back and presenting herself. After a gestation of nearly seven months, an unusually long period for a mammal of this size, the single young is born in May or June. Quills are well formed but not injurious to the mother, as the baby is born head first in a placental sac and the short quills are soft; quills harden within half an hour. The porcupine's life span is seven to eight years.

Predators other than the fisher include bobcats and coyotes. In some states, the porcupine is protected because it provides an easily obtained source of food to a person lost in the woods. A sharp blow on the nose with a stick kills it. Its quills, both natural and dyed, have been used for decorative purposes by native Americans, who also ate the animal's flesh.

△ A porcupine eating crabapple.

△ Baby porcupines do not have mature quills to defend themselves. They are all black and thus blend into the background rather than being obvious.

△ A close-up view of the sharp-pointed quills which have microscopic barbs at the tips. Quills embedded in flesh can be removed more easily if the ends are cut off, thus releasing air pressure.

NUTRIA

Myocastor coypus

The nutria is a South American rodent which has been widely introduced, especially in the Southeast, but also in Maryland, and southern New Jersey. The nutria is a large aquatic rodent with a long, scaly, sparsely-haired, rounded tail, and small ears and eyes. Like the muskrat, it lives in marshes, but may also be found in ponds and streams.

The hindfeet have the inner four toes webbed and longer than the forefeet. Males are larger than females. The muskrat is smaller, with a vertically flattened tail. The beaver is larger, with a large, horizontally flattened tail.

The nutria produces feeding platforms of aquatic vegetation and debris, five to six feet in diameter, which it occupies to rest and to avoid terrestrial predators. Also, it creates trails of flattened vegetation through the marsh. Nutrias may produce a chorus of pig-like grunts at dusk. Their elongated droppings may be found on the feeding platforms, paths or on the shore.

Myocastor means "mouse beaver". The species name *coypus* is Indian in origin and translates as "water sweeper." Nutria is a Spanish word for otter, but the web-footed nutria is a rodent and is not related to otters.

The nutria often feeds on land, but returns to the water when startled, often with a loud splash. It can remain under water for several minutes and often floats just under the surface with only eyes and nose exposed. It may dig its own burrow, use the burrow of other animals, or the lodge of a beaver or muskrat. It

will eat most terrestrial or aquatic green plants, and grain, sometimes dipping its food into water before eating. Like the lagomorphs (rabbits and hares) it re-ingests pellets in order to digest food more completely while at rest.

Courtship includes much chasing, fighting, and biting. After a gestation of about 130 days, a litter of one to eleven young (usually four to six) is born. The young are fully-haired and with eyes open. Litters are largest during periods of food abundance. The young swim with their mother and nibble green plants within 24 hours of birth.

THE HISTORY OF NUTRIA

In 1899, nutria were introduced for fur farming into California from South America, where they are native. The fur is used for coats and hats. But the fur farms failed due to the nutrias' destructive digging behavior and their need for huge amounts of food.

Nutrias eat up to 25% of their body weight in plant material every day. (Mink, in contrast, can thrive on small quantities of meat rendering by-products. Herbivores always require more food than carnivores.)

In the late 1930s promoters sold nutrias as aquatic weed cutters and for fur-raising "for fun and profit." Large numbers were imported to North America. It was hoped that they would be effective in clearing pond vegetation, but they went beyond this to eat useful plants as well.

With the failure of the fur farms and the weed clearing projects, many of the animals were released into the wild, and others escaped from inadequate pens. They were hard to confine in pens because of their ability to dig out, and the concrete needed to contain them properly was too expensive.

Feral populations are now established in at least 15 states in the US. Nutrias are highly gregarious, and wild nutrias are very tame. They don't seem to have much fear of man.

△ Nutrias, like other rodents, have lips that close behind their teeth so that they can harvest underwater vegetation without swallowing water. The mouth of the nutria in the photo above appears to be wide open, at first glance, but is actually tightly shut.

SIDE NIPPLES

The female nutria has four or five pairs of nipples located on the side of her torso, a unique adaptation which allows her to stand up to watch for predators while nursing young on the nest. No other North American mammal has mammary glands positioned in this manner.

THE NUTRIA FUR TRADE

The nutria is the number one fur-bearing animal in Louisiana; there is good demand for its belly fur. Most of the nutrias in Louisiana are descended from 20 individuals brought there in 1938. By the late 1950s there were an estimated 20 million nutrias in the state.

DOUBLE DIGESTION

Rabbits and certain other animals such as nutrias re-ingest their feces by eating them directly from the anus. The animals usually do this after returning to their nest at the end of their active period. There are two kinds of fecal pellets produced and only one kind is re-ingested. Re-ingestion is not merely an effort to derive more nutrition from the same food, but certain essential vitamins are created in the digestive tract through this process.

ANIMAL FUR

Hair, along with milk in mammary glands, is one of the key characteristics of mammals. All mammals have hair at some stage of their development (even if only a little bit such as the whales). No other vertebrates have hair. Some mammals have dense fur, while other mammals such as whales and elephants have only a few scattered bristles. The manatee has only a few hairs scattered over its body plus whiskers on its face.

Hair is composed primarily of the protein keratin, which is also found in fingernails, claws, hooves, and horns, and in the feathers of birds. Hair also contains the dark pigment melanin and other pigments.

Most mammals have three kinds of hair: facial whiskers which are sensitive to touch and which have sensing apparatus at the base; soft, fine hairs which make up the thick underfur of the animal; and longer guard hairs which protect this undercoat. The density of the hairs depends on the type of animal and the hairs' location on the animal. Sea otters have the thickest fur of any mammal, averaging 250,000 hairs (and up to 1,000,000 hairs) per square inch. In contrast, humans (who are not balding) have only 100,000 hairs on their heads, far fewer than the sea otter has on a single square inch of its fur.

Hair has a number of different functions on mammals. Insulation is the obvious function; a thick coat of fur helps prevent heat loss in winter. Many animals, such as white-tailed deer, grow different winter and summer coats. The winter coats are many times heavier and provide better heat retention.

Hair provides buoyancy to some aquatic mammals. Sea otters trap air bubbles in their hair so they can float more easily—oil from spills prevents this. The otters also spread a water-repellent oil from their skin to their hair.

Polar bears have hollow, translucent hairs. These hairs allow sunlight to reach their dark skin, and this helps them absorb heat. In zoos, their fur may appear green, because algae sometimes grow on their hair.

Some animals regulate their body heat by fluffing out or flattening their fur. Fluffed up fur retains heat better, while flattened fur lets heat escape. Goosebumps on people may have served to fluff up our fur, long ago, when we had real fur.

Some animals have markings on their fur as babies which they lose as they grow up. Panther cubs and deer fawn are spotted, for example, but the adults are not. Young animals are more vulnerable to predators, and spots or other markings probably help protect the youngsters by providing camouflage.

The quills of porcupines are modified hairs. Not only are they sharp and stiff, but they have scales that point toward the base, unlike most hairs, so after the quills penetrate a victims flesh, they are hard to pull out. Another example of modified hair is the rhino's horn which is composed of compressed hairs.

DOGS & FOXES

COYOTE
Canis latrans

The coyote is one of the most adaptable animals in North America. Its scientific name, *latrans*, means barking dog and the coyote does resemble a dog, such as a small German shepherd. Its common name comes from coyotl, the name used by Mexico's Nahuatl Indians. Coyote is pro-nounced *ki-o'-ti*.

The coyote was formerly a western desert animal, but has expanded its range greatly in recent years. Today it can be found from the middle of Alaska to Central America and throughout the eastern US. It inhabits many habitats, but it does best in scrub or brushy areas.

The coyote was not very common east of the Mississippi until the 1950s. It may have been responding to the availability of more open land in the East, but a major factor in its increase was probably the loss of the next larger species, the wolf, thus creating an open niche in some areas.

The coyote is the best runner among the canids. It can leap 14 feet. It normally cruises at 25 to 30 mph and up to 40 mph for short distances. Tagged coyotes have been known to travel distances of up to 400 miles. The coyote runs with its tail downwards at an angle, unlike dogs, which run with their tails curved upwards. The coyote is a strong swimmer and does not hesitate to enter water after prey.

Coyotes usually feed on rabbits, mice, ground squirrels, and other small mammals, birds, frogs, toads, snakes, insects, and many kinds of fruit, and carrion. Coyotes occasionally bring down larger prey, such as deer, with several individuals combining efforts, running

△ The sounds of the coyote are varied, but the most distinctive and best known is their howling, usually heard between dusk and dawn. It consists of a series of barks and yelps followed by a prolonged howl and ending with short, sharp yaps. Coyote howling keeps the pack, usually a family group, informed of the locations of its members and reunites them when separated. One call usually prompts others to join in, resulting in the familiar chorus heard at night throughout the west. It was seldom heard in the east when coyotes were rare, but is commonly heard now in many areas as coyote populations increase. Barking alone, with no howling, seems to be a threat display used to defend a den or a kill.

in relays to tire prey, or waiting in ambush while others chase prey into range. This happens most often in deep snow. Sometimes, a badger works with the coyote to catch smaller prey. While the badger digs for rodents at one entrance of a burrow, a coyote will wait at another to pounce on any prey that may emerge from an escape hole. Sometimes the badger gets the prey, sometimes the coyote. The coyote stalks like a pointer, freezing before it pounces.

Predators of coyotes once included mountain lions and wolves, but these are gone in the Northeast, leaving humans as their major enemy. Coyotes were particularly sought a few years ago when coyote pelts were quite valuable. Even today when pelts are of little value, coyotes are often shot on sight because of the idea, often mistaken, that they reduce native wildlife and prey on domestic livestock. Attacks on domestic livestock often prove to be the work of domestic dogs rather than coyotes. Numerous coyotes are killed on the highways. However, the coyote population continues to increase, despite trapping, poisoning, and motor vehicles.

Coyote dens are often in banks or on slopes. Den openings are often marked by a mound of earth and numerous radiating tracks. The typical den is a wide-mouthed tunnel, five to 30 feet long, with an enlarged nesting chamber at the end. The female may dig her own den, enlarge the burrow of a fox, or use a cave, log, or culvert. If the den area is disturbed, the female will move her pups to a new den.

Coyotes may pair for several years or life, especially when populations are low. Mating is from February to April, and the one to nine young (averaging about six in good rodent years, less in years with less rodents) are born usually in April or May in a crevice or underground burrow. Both parents, and sometimes individuals from earlier litters, help feed the young (regurgitated food when first weaned, solid food later).

Coyotes can often be seen moving swiftly across open country. Coyotes in the wild can usually be distinguished from dogs, even those dogs that look very much like coyotes. Dogs usually hold their tails in an upward arch, whereas coyotes hold their tails straight, but pointing downward at a 45 degree angle.

Coyote droppings look much like those of a dog but are full of hair. Coyote tracks likewise are similar to dog tracks, but are in a nearly straight line. There are four toes, all with claws. Tracks and droppings are often seen where runways intersect, or on a small hill or open spot where coyotes have stopped to watch for prey.

△ A coyote pouncing on its prey (which could be a bird or a rabbit).

△ The photo, above left, shows a baby coyote and the photo at right shows a coyote with its pup at the entrance to its den. Coyotes may make a round-about path to their den which helps to avoid leading predators to their young.

COYDOGS

Coydogs (hybrids of coyotes and domestic dogs) are usually larger than coyotes and usually lack the dark vertical line on the lower forelegs. Coydogs are thought to be abundant, but in reality they are relatively rare, especially now that coyotes have increased so dramatically. Hybridization is much more common when one species is rare and therefore has difficulty finding mates.

RED FOX
Vulpes vulpes

The red fox is a small, dog-like animal that behaves much like a cat. It has a long, bushy tail with a white tip. Color variations include a black phase (almost completely black), a silver phase (black with silver-tipped hairs), a cross phase (reddish-brown with a dark cross across shoulders), and intermediate phases. All have the white-tipped tail.

Red foxes live in mixed cultivated and wooded areas and brush lands.

Maternity dens are prepared shortly after mating which usually occurs in late January or February. The maternity den is usually in an area of sparse ground cover and often on a slight rise, providing a view of all approaches. The maternity den may also be in a stream bank, slope, or rock pile, and less frequently in a hollow tree or log. It is often the enlarged den of a woodchuck. Typically the entrance is a dirt mound with the main entrance up to one foot wide, slightly higher, often littered with leaves. There may be one to three less conspicuous and smaller escape holes. The den is usually marked by excavated earth, cache mounds where food is buried, holes where food has been dug up, and scraps of bones and feathers.

Red foxes form territories of about 250 acres.

Territories are marked with urine and caches are similarly marked also apparently to indicate "no food here."

After a gestation of about 51 to 53 days, one to ten kits (usually four to eight) are born March to May. When about one month old, they play above ground and feed on what is brought to them by their parents. At first, meat is predigested by their mother and regurgitated, but soon live prey is brought, enabling kits to practice killing. Kits disperse at four months, males up to 150 miles away or more, females usually less widely. Dens are abandoned by late August when families disperse. Adults also disperse, remaining solitary until the next breeding season.

The red fox is usually difficult to observe even when fairly abundant, as it is shy and nervous. It is primarily nocturnal, although it may be seen near dawn or dusk or on dark days.

It is omnivorous, feeding heavily on corn, berries, apples, cherries, grapes, acorns, and grasses. In winter, it feeds on birds and mammals, including mice, rabbits, squirrels, and woodchucks. Invertebrates such as grasshoppers, crickets, caterpillars, beetles, and crayfish comprise about a quarter of its diet. Food not eaten

at once is cached under snow, leaves, or soft dirt. Much of its food is of small mammals which it captures by pinning to the ground. It caches food by holding the item in its mouth while it digs with the front feet. The dirt is piled to the side and the item is put in the hole, covered with dirt, leaves, etc. Caches are relocated by memory and smell.

Adult red foxes rarely den up in winter. Rather they curl up into a ball in the open, wrapping the bushy tail about their nose and foot pads. At times, they may be completely blanketed with snow.

In the mid-eighteenth century, red foxes were imported from England and released in New York, New Jersey, Maryland, Delaware, and Virginia by landowners wishing to hunt with hounds. The gray fox had not yet expanded its range north into these areas. It was not a good substitute for the red anyway, as it cannot run as fast or as long. Red foxes of the Northeast now are probably combined strains derived from interbreeding of imported and native races, which, encouraged by settlement, gradually expanded their range south from Canada.

THE BOUNTY SYSTEM

For years, unregulated trapping and bounty payments took a heavy toll on foxes. The bounty system was an exercise in futility. It led to widespread unnecessary killing of many species, including red foxes, and probably caused some species to be eliminated from the Northeast. Individual problem animals would have been killed anyway without the bounty, and the system was filled with fraud. Fortunately, most bounty payments have been abolished. As a result, foxes are now doing much better, although competition with the coyote, which is also expanding its range, may have a restraining effect.

△ A red fox at the entrance to its den. The young can be seen inside.

GRAY FOX

Urocyon cinereoargenteus

Gray and red foxes are often misidentified because the gray fox has quite a bit of red on its body. However, the red fox is much redder and has a white tail tip. The tip of the tail in the gray fox is not white.

The gray fox is grizzled gray above, reddish below and on the back of the head. The tail has a black mane above, and a black tip. The throat is dark.

It lives in varied habitats, but the gray fox is associated much more with wooded and brushy areas than are red foxes. Gray foxes mark trees and other scent posts with urine. This is often made noticeable by spattered urine stains and melting of snow. The tracks, usually in a straight line, are similar to those of a very large domestic cat, except that non-retractile claws may show. The tracks are also very similar to those of the red fox, but are often smaller with larger and more sharply defined toes because of less hair around the pads.

Gray fox dens are usually in natural cavities, thus the entrance size varies greatly. Favored den sites are in woodlands and among boulders on the slopes of rocky ridges. Gray foxes will dig a den if necessary but they usually den in clefts, small caves, rock piles, slash piles, hollow logs, and hollow trees, especially oaks. Sometimes they enlarge a woodchuck burrow. The entrance is occasionally marked by snagged hair or a few telltale bone scraps and there are usually several auxiliary or escape dens nearby. In rare instances their burrows are marked with conspicuous mounds like those of the red fox.

Mating is in February and March. The two to seven young (usually three or four), are born in March or April. They are weaned at three months, and begin hunting for themselves at

△ Gray foxes use trees as resting places. Red foxes do not climb trees.

BK

△ This gray fox mother is nursing her young offspring. They may venture out of the den to play at about three to four weeks, but only under the watchful eye of the mother.

four months. The male helps tend the young, by bringing food, but he does not den with them.

The gray fox is primarily nocturnal, but is sometimes seen foraging by day in brush, thick foliage, or timber. This is the only American canid with true climbing ability. It occasionally forages in trees and frequently takes refuge in them, especially in leaning or thickly branched ones.

Unlike the red fox, the gray uses dens all winter. It growls, barks, or yaps, but less frequently than the red fox. Omnivorous, it feeds heavily on cottontail rabbits, mice, voles, other small mammals, birds, insects, much plant material, including corn, apples, persimmons, nuts, cherries, grapes, pokeweed fruit, grass, and blackberries. The most important predators are domestic and wild dogs, bobcats where abundant, and man.

GS

Caches of food are in heaped or loosened dirt, moss, or turf, frequently paler than surrounding ground. Excavated cache holes are shallow and wide, since foxes seldom bury very small prey except near a den in whelping season.

The droppings are small, narrow, roughly cylindrical, and usually sharply tapered at one end. Because gray foxes eat more berries than red foxes, their stools are often darker, particularly where wild cherries abound.

BEARS

Family Ursidae

The three species of North American bears (the black, the grizzly, and the polar bear) are the largest of all terrestrial carnivores, ranging up to 1,700 pounds. They have five claws on each foot and, like man, walk on the entire sole with the heel touching the ground. They have powerful, densely furred bodies; small, rounded ears; and small eyes set close together. Their vision is poor, but their sense of smell is keen. Bears are highly omnivorous, eating leaves, twigs, berries, fruit, and insects as well as small mammals.

Although most people believe that bears hibernate, it is not precisely true. They enter a protected area and sleep away the harshest part of winter, but their sleep is not as deep as in true hibernators. In true hibernators, body temperatures approximate those of the environment, often near freezing. Temperatures of bears fall only a few degrees below normal.

The young are born while the female is denned up. Although the eggs are fertilized when the female mates in late spring or early summer, six or seven months may pass before the embryos become implanted in the uterine wall, after which they develop rapidly. When born, bears are the size of rats, generally weighing only one-half to one pound, which makes the magnitude of their eventual growth greater than that of all other mammals except marsupials. North American bears produce a litter every other year.

All bears are dangerous, but especially when accompanied by cubs, surprised by the sudden appearance of humans, approached while guarding kill or feeding, fishing, hungry, injured, breeding, or when familiarity has diminished their fear of man.

Once distributed over much of North America, bears have been eliminated from most areas that have substantial populations of humans.

BLACK BEAR

Ursus americanus

The eastern black bear is almost completely black. The tail is very short, three to seven inches. Males are much larger than females. They can weigh close to six hundred pounds, but are usually much less. In the area covered by this book, bears occur in New England, New York, and Pennsylvania south through the Appalachians. Habitat in the Northeast is primarily forests and swamps.

Where black bears occur, feeding signs are common; logs or stones turned over for insects; decayed stumps or logs torn apart for grubs; ground pawed up for roots; anthills or rodent burrows excavated; berry patches torn up; fruit-tree branches broken; and rejected bits of carrion or large prey, such as pieces of skin, often with head or feet attached. Also, bear trees may be scarred with tooth marks, often as high as a bear can reach when standing on its hind legs. The longer claw slashes can be seen even higher on the tree. They are usually diagonal but sometimes vertical or horizontal. In spring, furrowed or shaggy-barked trees are used repeatedly, sometimes by several bears, as shedding posts to rub away loose hair and relieve itching. These show rub marks and snagged hair.

Droppings are usually similar to a dog's, dark brown, roughly cylindrical, sometimes coiled. They often show animal hair, insect parts, fruit seeds, grasses, root fibers, or nutshell fragments. Where bears have fed heavily on berries, their droppings may be a liquid black mass.

Where generations of bears have used trails, the trails are well worn, undulating, and marked with depressions. Tracks of bears are very large. Individual hind prints look as if they were made by a flat-footed man in moccasins, except that the small-

EPB

est toe is innermost and occasionally fails to register, and the large toe is outermost.

Mating is in June or early July. The litter of one to five young (usually two or three) is born in January or early February, generally every other year. The young weigh only a half-pound at birth. Females mate in their third year, with most producing one cub the first winter, two at the next breeding. The young are born while the mother sleeps in the den, and the nearly naked newborns nestle into her fur. The mother often lies on her back or side to nurse, but sometimes sits on her haunches, with the cubs perched on her lap, almost like human infants.

The black bear is primarily nocturnal, but may be seen at any time, day or night. It ranges in a home area of eight to ten square miles (sometimes as many as 15). Bears are solitary except during the mating season and when congregating to feed at dumps. They are clumsy when they walk, but can attain surprising speed when they trot, with bursts of up to 30 mph. Powerful swimmers, they also climb trees, either for protection or food.

While small to medium-sized mammals or other vertebrates are eaten, most of the diet consists of vegetation, including twigs, buds, leaves, nuts, roots, various fruit, corn, berries, and newly sprouted plants. In spring, the bear peels off tree bark to get at the inner, or cambium, layer. It tears apart rotting logs for grubs, beetles, crickets, and ants. A good fisherman, the black bear often wades in streams or lakes, snag-ging fish with its jaws or pinning them with a paw. It rips open bee trees to feast on honey, honeycomb, bees, and larvae.

In the fall, the bear puts on a good supply of fat, then holes up for the winter in a sheltered place, such as a cave, crevice, hollow tree or log, or roots of a fallen tree. Excrement is never found in the den. The bear stops eating a few days before retiring to the winter den, but then consumes roughage, such as leaves, pine needles, and bits of its own hair. These pass through the digestive system and form an anal plug, up to one foot long, which is voided when the bear emerges in the spring.

Bear hunting is a popular sport. The hides are sometimes used for rugs.

△ Bears become dangerous as they lose their fear of man, especially in parks. People have been killed by bears in a few unfortunate cases. Bears should never be regarded as harmless. Although they seem cute sometimes, it is best not to feed them because it encourages them to come in dangerously close contact with humans.

RACCOON *Procyon lotor*

The raccoon is sometimes called a "masked bandit" because of its facial markings. However, it is most often called by its slang name, "coon."

The raccoon is a medium sized mammal with a bushy tail with four to six alternating light and dark rings, and with a black mask outlined in white. The animal is reddish-brown above, with much black and grayish below. The ears are relatively small. Raccons usually weigh about 12 to 30 pounds, but an occasional animal weighs over 50 pounds. The raccoon lives only in the New World where it occupies various habitats. It is most common in wooded areas containing streams.

The most distinctive signs of raccoon are its many tracks most often along the waters edge (or where the water has dried up). Often parts of crayfish, mussels, and fish occur there also. Broken stalks, shredded husks, scattered kernels, and gnawed cob are often found in cornfields, although these signs could also indicate that groundhogs or squirrels or some other creature has been at work.

The den is usually in a hollow tree, which may have its base and trunk scratched or bark torn, and scat accumulated about its base. Droppings are inconsistent in shape, but are generally cylindrical, uniform in diameter, and about two inches long. They resemble, and are difficult or impossible to distinguish from those of opossum and skunk, but are often deposited on large tree limbs, stones in a stream, or logs.

Hind prints are three and a quarter inches to four and a quarter inches long. They are much longer than wide and resemble a miniature human footprint with abnormally long toes. The fore print is much shorter, about three inches long and

EPB

almost as wide as long. Claws show on all five toes. Tracks are large for the animal's size because the raccoon is flat-footed. Its stride is six to 20 inches, averaging 14 inches. When walking, the left hind foot is almost beside the right forefoot. When running, it makes many short, lumbering bounds, bringing the hind feet down ahead of the fore feet in a pattern like oversized squirrel tracks.

Mating in the north is usually in February. The female accepts only one male per season, and males travel miles in search of mates. The male remains in the female's den a week or more, then seeks another mate. The female most often makes a leaf nest in a large hollow tree but other protected places may be used such as culverts, caves, and rock clefts, woodchuck dens, or under wind-thrown trees.

A litter of one to seven young (usually four or five) is born in April or May. The young weigh about two ounces at birth. They open their eyes at about three weeks, clamber about the den opening at seven to eight weeks, and are weaned by late summer. At first, the mother carries them about by the nape of the neck, as a cat carries kittens, but soon leads them on cautious foraging expeditions, boosting them up trees if threatened, but attacking ferociously if cornered.

Some young disperse in autumn, while others may remain for a time, but are driven away by the female before she bears her next litter, as den space is limited. Some yearling females may be pregnant themselves and must quickly find dens of their own.

The raccoon is nocturnal and solitary except during the breeding period. Although territories overlap, when two meet, they growl, lower their heads, bare their teeth, and flatten their ears. The fur on the backs of their necks and shoulders stands on end, generally with the result that both back off.

During particularly cold spells, the raccoon may sleep for several days at a time but does not hibernate.

Raccoons have a number of vocalizations, including purrs, whimpers, snarls, growls, hisses, screams, and whinnies.

Omnivorous, they eat grapes, nuts, grubs, crickets, grasshoppers, voles, deer mice, squirrels, other small mammals, and birds' eggs and nestlings. They spend much time foraging along streams and swim (but do not dive) for crayfish, frogs, worms, fish, dragonfly larvae, clams, turtles, and turtle eggs. They may raid muskrat houses to eat the young. They will climb trees to cut or knock down acorns as do squirrels. In residential areas, raccoons often tip over or climb into garbage cans to forage.

The raccoon's nimble fingers, almost as deft as a monkey's, can manipulate food and can even turn doorknobs and open refrigerators. The raccoon has long been said to wash its food. In fact, its scientific name, *lotor,* means a washer. The objective, however, is not to clean the food, but rather to knead and tear at it, feeling for matter that should be rejected. Wetting the paws enhances the sense of touch. The common name of the raccoon comes from *aroughcoune,* used in colonial times by the Algonquin Indians of Virginia to mean "he scratches with his hands."

Raccoon hunting is popular in late autumn, when raccoons are very active, fattening themselves for winter. It is usually done at night. Dogs trail the raccoon until it trees; then hunters follow the baying of the hounds to the tree. They may shoot the animal, or if the raccoon is small, the hunters may spare it. For many, the sport lies in listening to their hounds and observing the skill of their performance. Sometimes, instead of treeing, a raccoon leads hounds to a stream or lake. A dog that swims well can easily overtake a raccoon in the water, but the raccoon, can usually beat a single dog in a fight.

Pelts, at times, have been valuable, and during the 1920s, coonskin coats were a collegiate craze.

RACCOONS AS PETS

Even though baby coons are cute and gentle, they do not make good pets. When they grow up, they may become extremely aggressive and uncontrollable. Keeping wild animals as pets often causes problems. Keeping a raccoon as a pet may result in bites, the destruction of property by males reaching maturity or even the death of a child. Friends or neighbors bitten by a pet coon have to worry about rabies.

Raccoons were so popular at one time they were even kept as pets in the White House. Grace Coolidge, wife of President Calvin Coolidge, had a raccoon named Rebecca, and President Herbert Hoover had a raccoon named Susie. Each lived in the White House for the duration of that president's terms. It's no accident that both these raccoons were females. If they had been males and allowed to run free, they might have caused considerable damage to the White House unless they had been neutered. Today, pet raccoons in the White House would probably be viewed unfavorably by large numbers of the voting public.

A DISTINCTIVE HUMPED POSTURE

Note the distinctive peaked shape of the coon's back, with the head low and rump high. This is particularly noticeable when the animal is walking or running. The reason is that the raccoon's hind legs are much longer than its front legs. Also, the shape of the spine is a factor.

Raccoons often stand on their hind legs while using their front paws to manipulate food. Well developed hind limbs may be associated with their climbing ability.

When walking, the coon holds its tail lower than many other similar-sized animals.

Raccoons are alert and curious, traits celebrated in native American folklore. After learning how to open a container, a raccoon can remember how to do it for as long as a year without practicing. In addition to having a good memory, raccoons have considerable ability in solving problems. They do at least as well as a cat, and sometimes even better than apes and monkeys.

Raccoons climb easily. They can jump from heights and land on their feet like cats. Raccoons can live as long as sixteen years. However, only one in a hundred lives as long as seven years in the wild, and most raccoons do not survive beyond two years. The main cause of death in juveniles is lack of food. Baby coons are weaned at a very young age. When not well fed, they become vulnerable to intestinal parasites. Automobiles are probably the greatest cause of death among adults.

RACCOON BREEDING HABITS

Raccoons do not live together as mated pairs. The males mate with several females. During the breeding season, females find a den. The male raccoon locates a female and, if she is willing, moves into her den for a short period of mating. Afterwards, the male resumes his wandering lifestyle.

When the young are born, the female cares for them and defends them. Raccoon mothers are unlikely to abandon their young even in the face of great danger.

The raccoon mother leaves the den only long enough to obtain food. She does not receive any help from the father, and in fact, would not accept any contact with a male raccoon during this period.

Raccoons are born with fur but are very helpless. Their eyes are closed at birth and do not open for several weeks. The young don't even leave the den until they are eight to twelve weeks old. Raising young coons requires a great deal of effort by their mother.

Juvenile raccoons vocalize by chattering. Adults make similar sounds, but in deeper voices, including growls and barks.

THE FAMILY PROCYONIDAE

Raccoons are members of a family which includes ringtails, coatimundis, and kinkajous. These animals all have long tails with dark and light banding. The cheek teeth are blunt, indicating that they eat a wide variety of foods, including vegetable matter. They all have five clawed toes on each foot and, except for the ringtail, they walk flat on the soles of their feet as do bears and humans.

FEEDING COONS

It is unwise to feed raccoons because they can do lots of property damage, such as damaging buildings and digging up new plantings. Feeding them brings an abnormally high population of raccoons into the area, and they may tear at screens to get into houses for food. They are also likely to get into territorial battles. And there's always the threat that they might be carrying rabies. Most important, however, is that raccoons carry a roundworm, Baylisascaris, that is dangerous to humans and other animals.

NOT A GOOD IDEA!

It is a myth that raccoons always wash their food before eating it. The raccoon's species name, lotor, means washer, because raccoons were once thought to wash their food, but this is incorrect. they are merely kneeding it to get rid of inedible pieces. Raccoons do spend much of their lives around ponds and streams, but most of the photos of raccoons holding food over water actually show raccoons harvesting food from the water. Raccoons obtain much of their food by using their paws to search underwater around rocks and crevices for small fish, frogs, and crayfish. Raccoons have considerable ability to identify, hold, and manipulate these food items with their dexterous front paws. Another related myth is that raccoons must wet their food because they lack salivary glands, but this is simply a folktale.

ANIMAL SIGNS

Wild animals are shy and elusive creatures. Many are active only at night, live underground, or live in dense vegetation. Even those animals that live near humans, such as raccoons and opossums, prefer to remain undetected. Of all the species of mammals, how many are commonly seen? Gray and fox squirrels are active during daylight and are conspicuous. Chipmunks and red squirrels are often obvious. Deer and rabbits can often be seen by the side of the road. The eye-shine of a fox or other animal is sometimes seen while driving down a narrow road at night. But most wild animals are rarely encountered.

However, mammals and other animals leave signs of their presence which are often more likely to be seen than the animals themselves. The most common signs of animals are their tracks, but other clues to their presence include runways, burrows, nests, mounds, lodges, scrapes, and scat (feces).

Tracks: Tracks left in sand or in mud along a creek are often obvious, and most mammals, at least the medium to large mammals, are identifiable by their tracks. Tracks indicate how large the animal is, how many toes are present, the size of the front foot versus the back foot, features that when considered together often reveal the identity of the animal.

Felines (cats) and canines (dogs and foxes) can be of the same size and have similar footprints. Both walk on the balls of their feet, so both have a very short track, and there is no impression of a heel. The main difference is that cats have retractable claws (which helps keep their claws sharp) and dogs do not. A dog print will often show the impression of a claw, while the footprint of a cat would not. By comparison, bears and raccoons walk flat on their feet like humans, so their tracks show a heel impression for the hind foot.

Deer, cows, goats, pigs, and horses walk on their toes (or hooves). As with the felines and canines mentioned above, this lifting of the heel is an adaptation for running which effectively lengthens the leg. Deer have two toes on each foot and their tracks are different than those of horses which have only a single toe. There are differences between tracks of different species that can help make an identification (provided it is a good track), but it is not always easy. Tracks are often not clear. A track in sand looks different than a track in mud. It takes a great amount of time to become skilled in identifying tracks.

Scat: Scats (feces) from different animals vary in shape, size, and content, and also in the manner of deposition. Cats mark their territory with feces, urine, and scent, and often attempt to cover it by scraping soil and leaves with their back feet, leaving a rectangular swath called a "scrape." Otters deposit their feces in communal latrines or on high ground as a means of advertising their presence. These deposits may also be accompanied by scrapes. Biologists often collect the scat of animals to analyze the contents and learn what the animals have been eating. This provides valuable information about the community of animals present in a specific habitat as well as the nutritional needs of the larger animals. Seeds, hair, bone fish, scales, shells and insect hard parts come through the digestive track relatively unchanged and can often be identified to the species level. However, scat analysis often gives skewed results because soft parts are absent or underrepresented as they digest faster than hard parts.

Owl pellets are the dried, regurgitated bone and hair balls which represent the indigestible parts of an animal eaten by an owl. They are often found beneath the tree where an owl has been feeding. Although they are not produced by a mammal, they can reveal much about the small mammals present. Over time, skulls from owl pellets can track changes in the community of animals. For example, it can signal the introduction of an exotic species or an invasion of Norway or black rats. When the pellets are pulled apart, they reveal skulls, limb bones, and hair, all of which help reveal what kind of animal was eaten.

Other signs: Many small animals create pathways through vegetation. The width and location of these runways helps to determine what kind of animal uses them. There are recognizable differences in the mounds made by moles, and certain insects. There are a variety of nests, such as the globular nest of the golden mouse, or lodges probably made by muskrats or beavers. Other signs of animal activity include burrows, signs of rooting by feral hogs, trees chewed by beavers, nuts eaten by squirrels, and grass cuttings made by voles.

None of this is as thrilling as seeing a live animal in the wild. But a walk in the woods can be much more interesting with a little knowledge of the signs animals leave behind.

RABIES

Rabies was one of the first diseases known to be transmitted from animals to man. Rabies is a Latin word derived from an old Sanskrit word which means "to do violence." The first reference to rabies appeared in the Samarian code at about 1885 B.C. which mentioned "...mad dogs which bite people causing death."

Rabies is a viral nerve disease which generally begins with the bite of an infected animal. After the bite occurs, the virus moves up the neurons (the nerve cells) toward the brain. Usually, no symptoms are noticed during this incubation period until the virus infects certain areas of the brain. Then the clinical signs appear and shortly thereafter the person or animal dies. Death usually occurs from paralysis of the diaphragm, which stops respiration. About the same time that it reaches the brain, the virus also moves into the salivary glands, and thereafter it can be transmitted by saliva, through bites.

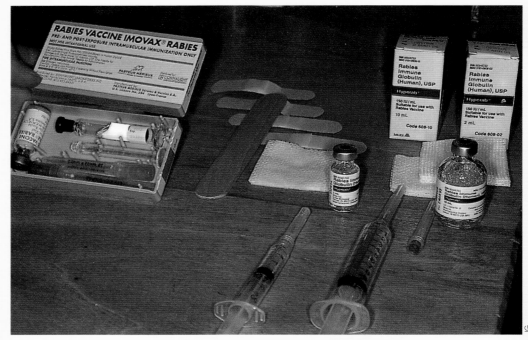

△ Complete set of instruments and vaccines for the treatment of bites by animals suspected of being rabid.

Different individuals respond to the virus in different ways and show different symptoms. Fear of water is a common symptom (for this reason rabies is also known as hydrophobia). But fear of water is not always present. Rabies is almost always fatal in humans once the clinical signs appear. There are only three documented cases of survival.

In the US, rabies was first observed in Virginia in 1753. By 1990 it had been recorded throughout the country.

Rabies in dogs and cats, people, and wildlife are all interrelated. Rabies passing from person to person is almost nonexistent. People usually get rabies only as a result of being bitten by a rabid animal.

Up until the 1950s, humans acquired rabies mostly from dogs, but the availability of a rabies vaccine for dogs completely changed the situation. Louis Pasteur, of pasteurized milk fame, is the inventor of the rabies vaccine for dogs. Starting in the late 1940s and early 1950s, local governments in the United States made a great effort to vaccinate dogs for rabies. The lack of human cases from dog bites in the US in recent years is a tribute to the success of this campaign.

Since most dogs have been vaccinated, the greatest danger of rabies today is from wild animals, although people often do not vaccinate pet cats, which may encounter rabid wild animals. Most cases arise from the bites of raccoons, foxes, bats, and skunks. Wild animals, including wild animal pets, cannot be vaccinated, because there is no way to vaccinate them, and because there is no licensed vaccine for these animals in the US. This is one of many good reasons for not keeping wild animals as pets.

Rabies is generally passed via infected saliva, when people or animals are bitten by infected animals. However, an animal with rabies which does not yet have the virus in its saliva cannot transmit the disease through its bite. In general, only infected animals showing symptoms of rabies have the virus in their saliva and are capable of transferring the disease, but there are exceptions. A skunk may have the virus in its saliva for months before it shows any signs of rabies.

Although virtually all rabies cases are acquired through saliva, there have been some unusual cases in which skunks passed the virus through the mother's milk. Rabies has been acquired by inhaling the air in bat caves, but this mode of transmission takes place only through prolonged exposure.

Most animals tested are near populated areas, where the risk to pets and humans is greatest, so the actual incidence of rabies in the wild is unknown. As with people, a wild mammal rarely recovers once the rabies symptoms have begun.

The time from exposure, or being bitten by a rabid animal, until the disease appears can be up to two years. Each species of animal, including man, has a different incubation period. The average incubation period for man is 21-28 days in the case of "street" rabies and much longer in the case of bat rabies. However, the length of the incubation period also depends on the location of the bite. If the bite is on the face or head, the disease may appear within a few weeks. If a toe is bitten, it is far less likely that the disease will appear at all and if it does, it might take six months to a year. Because of its long incubation period, rabies is the only virus disease presently known which can be prevented by a vaccine taken after exposure.

There are two types of rabies: a dumb form (primarily in bats) and a furious form, or "street" rabies (primarily in other animals). The victims of the furious form, mostly dogs, often exhibit biting and snapping behavior and become very aggressive. Hydrophobia, a fear of water, is also present. The dumb form is the form most common to cattle. The victims of the dumb form rarely attack, but they behave abnormally. A rabid fox or raccoon, for example, may walk about in daylight, a very uncharacteristic behavior for these nocturnal animals.

All mammals can get rabies, but the susceptibility of species varies enormously. Foxes are very susceptible to rabies, followed by skunks, cats, raccoons, and bats, and then cattle, man, horses, and dogs. In many areas of the US, skunks show the highest incidence of rabies among the animals tested.

Some mammals, such as opossums, seem highly resistant to rabies, or have not yet been documented as vulnerable. Rabies also appears to be very uncommon in rodents. Since 1957, thousands of squirrels, rats, and mice have been examined, and only one was found to be rabid. There is no known case of a person getting rabies from a rodent's bite. Rabies has never been documented in birds.

The dumb form of rabies found in bats is quite different from the "street" form. The sick bats just lie still. The rate of infection found in bat populations is very low, far below one percent in normally acting bats.

WHAT TO DO IF A WILD ANIMAL BITES YOU?

In the US, 20,000 people are treated each year for possible exposure to rabies. The injections are no longer a painful ordeal for the patient as they used to be.

If a wild animal is suspected of having rabies, and it has bitten someone or caused alarm in some way, the first thing to do is to call Animal Control so that they can kill or capture the animal. If the animal has to be killed to be captured, the head must not be destroyed, because the brain tissue has to be examined. Anyone touching the animal should use gloves to avoid contact with its saliva. If the animal is a bat, it should be killed and frozen, or at least kept cool.

When diagnosing rabies in animals, the brain tissue is examined, so the animal must be sacrificed. A pathological examination is conducted to determine whether or not the animal had rabies by examining the brain for black spots (negri bodies). Bat brain tissue must be examined using fluorescive dye as negri bodies do not show in a slice of bat brain tissue. If it is determined that the animal which has bitten a person has rabies, or if the animal cannot be located, treatment with rabies vaccine should be started immediately.

Veterinarians and other people that may be exposed to rabid animals, such as wildlife officers and animal control officers, are periodically vaccinated against rabies as a precautionary measure

▷ The series of rabies vaccine injections is now given in the arm rather than in the stomach, as was the practice in the past. The treatment now is not nearly so difficult or painful as it was years ago.

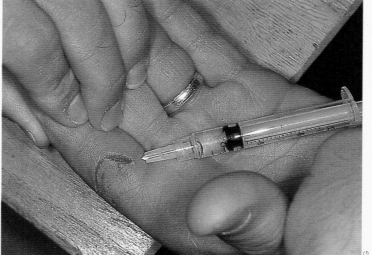

◁ One of the first steps in the modern treatment of bites from suspected rabies carriers is to inject the site of the wound with the rabies serum. The serum works immediately to start killing the virus. This is important because the vaccine takes up to 60 days to build a person's immunity.

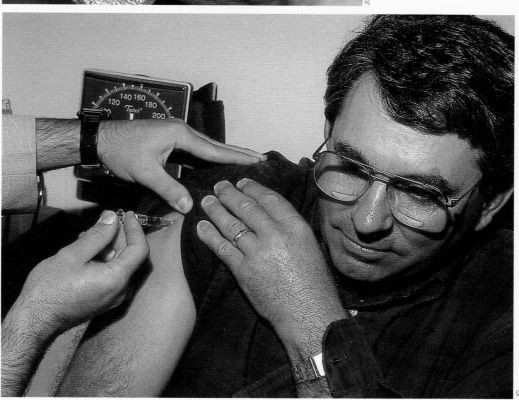

RABIES IN THE RECENT PAST

The following historical account gives a very vivid description, in the colorful language of that period, of the course of "hydrophobia," or rabies, in humans. Dr. Lancaster wrote a letter to the editor in 1894, reporting about a case he had attended:

"Having recently been called up to treat a case of hydrophobia, rabies, ... On the 27th of last July, Margaret Anderson, a woman, age 57 years, was bitten by a sick cat. The animal's teeth went into the fleshy part of the hand and holding on until pried lose. Four days later I was called and found an extensive cellulitis on the hand which I incised, liberating pus. I ordered the hand poulticized with carbonized flaxseed and ... in the course of time the hand returned to its normal condition. On the 19th of November, I was called again and found the patient in a nervous condition with an ill-defined dread manifesting itself with a constant change of positioning and in the appealing look and words of the patient to do something for her relief. She complained of a stiffening of the hand and arm accompanied by shooting pains extending from the hand to the shoulder. I at once recognized the symptoms of hydrophobia. I offered water to the patient. However, there was some difficulty in swallowing. This symptom increased until the sight or even mention of water brought on painful laryngeal contractions with the expression of fear and pain and the patient longed to have it removed from sight. There was an indescribable wild, scared look about the patient with constant restlessness ... The attempt to give .. other remedies brought on such paroxysms of fear and pain that I gave her no remedies other than hypodermic injections of morphine and atropia.

This I gave for controlling the suffering. The patient died on the 20th, some 48 hours after the first decided symptoms. At no time did the patient attempt to harm her attendants or was there any attempt to mimic the bark of a dog or cry of a cat. Though there was a frequent reference to the cat bite by the patient who clearly recognized her trouble and the unfavorable prognosis. There was a constant dripping from her mouth of ... saliva, obstinate constipation and inability to swallow liquids or solids the last 24 hours of life."

Note: Dr. Lancaster's method of diagnosis (offering the patient water and observing her distress) was commonly practiced many years ago. However it is now known that different individuals respond to the virus in different ways and show different symptoms. Fear of water is not always present.

THE WEASEL FAMILY

WEASELS, MARTENS, FISHERS, MINKS, BADGERS, AND OTTERS
Family Mustelidae

The family Mustelidae is greatly variable. It includes, for example, the arboreal marten, the aquatic otters, and the burrowing badger. Mustelids are small to medium-sized animals, with long, low-slung bodies, short legs, and short rounded ears. All are solitary, primarily nocturnal, and active throughout the year.

Until 1977, skunks were classified with the Mustelidae, but as of now, they are placed in their own family, the Mephitidae. Like skunks, most mustelids have paired anal scent glands. These are highly developed in skunks for defense, whereas in mustelids, the secretions are used more as social and sexual signals.

Many mustelids exhibit delayed implantation. The fertilized eggs undergo cell division to the blastula stage (hollow ball of cells), and floats free in the uterus for an extended period instead of immediately implanting in the uterine wall. Development of the embryo is suspended until implantation finally occurs. This delay is advantageous, allowing animals to mate in summer or autumn and bear young in spring when food is plentiful and conditions for growth and survival are optimal.

Many mustelids have a poor public image. Weasels and their kin are mainly thought of as killers of poultry and nesting game birds, although modern poultry-raising methods have minimized such predation. In fact, these carnivores serve to control rodents. On the other hand, mustelids have thick silky coats that make them valuable fur-bearers.

MARTEN
Martes americana

The marten is a weasel-like animal, but larger than a weasel, with an orange or buff throat patch, and a long, bushy tail. Its color varies from dark brown to blond. Males are larger than females. The fisher is much larger than the marten and lacks the orange throat patch. The mink is darker, with a shorter tail and white on the chin, and it too lacks the orange throat patch. Martens usually occur in coniferous forest.

Mating occurs in midsummer, followed by delayed implantation. Two to four young are born in a leaf nest, a tree hole, or fallen log about April.

The red-backed vole is the main food of the marten thoughout its range, but it also eats other mice, red squirrels, flying squirrels, rabbits, and birds. Carrion, eggs, berries, conifer seeds, and honey are also eaten. Martens pounce on their prey. Surplus meat is sometimes buried, although martens are generally poor diggers.

They are inquisitive and can be coaxed from their dens by making squeaking noises. Their pelts are valuable, thus over-trapping has led to their extirpation in many areas, and lumbering has destroyed much of their habitat and reduced populations in other areas. However, martens are protected and are presently making comebacks in many localities.

Martens spend much of their time in trees. They are active in early morning, late afternoon, and on overcast days. Their home range is five to fifteen square miles. Both sexes establish scent posts by rubbing their scent glands on branches.

Martens are solitary and avoid each other. If two meet, they bare their teeth and snarl.

FISHER

Martes pennanti

The fisher is a long, thin dark brown animal, with a bushy tail. The marten is similar but has an orange throat patch, and the mink is much smaller and has a white chin patch. In the Northeast, the fisher is found in northern New England and New York in mature dense forests. Fishers avoid open areas.

This is the only mammal that regularly kills porcupines, and the droppings may contain porcupine quills, as well as fur, bone, berries and nuts. Fisher tracks are similar to those of other mustelids, but are larger than most, and have the claws showing. Often the trail ends when the fisher climbs a tree.

Mating is in March and April, right after the female gives birth to her one to five young in a rock crevice or hollow tree. Total gestation is nearly one year, including a ten month delay in implantation.

Like most mustelids, the fisher is primarily nocturnal, although it is sometimes out during the day. Snowshoe hares and porcupines are important prey of the fisher, but it also eats squirrels, mice, chipmunks, carrion, fruit and plants.

Few porcupines are actually killed by fishers, as they are difficult to kill. However, the fisher has no competition for porcupine, and porcupines provide a great deal of food. Porcupines are killed only on the ground. The fisher circles, waiting its chance for attack, while the porcupine is turning to keep its back to the fisher. The porcupine will also put its face against an object such as a tree, as the fisher kills by repeated bites at the face. The fisher may climb the tree, then come down to force the porcupine away from the tree. The porcupine is finally killed after being weakened by repeated facial wounds. Internal organs are eaten first, and the porcupine will last the fisher for two or three days. Although fishers are adapted for killing porcupines, occasionally one is injured or even killed by the quills.

Red squirrels are often eaten. They are usually caught on the ground. The amount of food needed by a fisher is about one snowshoe hare per week, a red squirrel or two, or up to about 22 mice per day.

Although the origin of its common name is unknown, the mink's habit of fishing may have been mistakenly ascribed to the fisher. It is a good climber and swimmer. Its home range is 50 to 150 square miles, and even larger in winter when food is scarce. It primarily stays on the ground, where it uses well-established trails and runs on fallen logs. However, it can climb and moves from branch to branch in

conifer trees.

The fisher has maternity dens and also temporary dens. Maternity dens are in hollow trees, but the fisher makes temporary dens in hollow trees, rocky crevices, underbrush, or holes dug in snow.

If disturbed, the fisher hunches its back like a cat and may hiss, growl, snarl, or spit. The fisher is a valuable fur-bearer. Female skins are especially prized. In many areas it has been extirpated primarily because of trapping and loss of habitat.

SHORT-TAILED WEASEL OR ERMINE

Mustela erminea

The short-tailed weasel, or ermine, is a tiny carnivore which is primarily a forest-dweller, but which may be found on open lands and farms. In the northern part of its range, it turns white in winter except for the black tail tip, nose, and eyes. In the southernmost parts of its range the short-tailed weasel does not turn white in winter. The winter-white fur of both the ermine and long-tailed weasel is known as ermine and is highly valued.

Males and females are separate most of the year. Males have much larger territories than females and they encompass the ranges of several females. The male will mate with the female and even her six-week old female offspring. Sometimes the eyes of the young may not even be open. The male grabs the female by the scruff of the neck and drags her around. Copulation lasts from two to twenty minutes and is repeated often. There is rapid turnover of males on the territory, so the female may mate with several males. The short-tailed weasel builds its nest in a protected area, such as under a log, rock pile, or stump.

The short-tailed weasel hunts mainly on the ground, but it can climb trees and occasionally pursues prey into water. It alternates short active periods (10 to 45 minutes) with long rest periods (three to five hours) throughout the day and night. A curious animal, it may often be attracted to a human observer by

△ This short-tailed weasel is changing from winter white to spring brown.

△ This is the rich brown summer color of the short-tailed weasel.

a series of squeaks. Its main food is mice and voles, chipmunks, and shrews, but it also eats baby rabbits, birds, frogs, lizards, small snakes and even insects.

This species pounces on its vertebrate prey with all four feet, biting through the neck near the base of the skull. It sometimes attacks something larger than itself, even several times its own weight, but only when an especially good opportunity arises, or in times of scarcity, for this behavior is very dangerous, and can lead to its own death. After making a kill, weasels lick

the blood from their prey first, giving rise to the false notion that weasels suck the blood of their prey. Also, weasels kill all available prey, storing what they can't immediately use. This behavior gives an advantage to an animal whose meals are not always assured, but has given rise to the notion that weasels are bloodthirsty.

This species has many predators; hawks, owls, and predaceous mammals. Probably only a few individuals live more than two years.

LEAST WEASEL
Mustela nivalis

The least weasel is a tiny weasel with an inch-long tail. Its total length is only about seven or eight inches. It turns white in the northern part of its range, but most individuals in the northeastern US do not. Its weight is less than two ounces. The short-tailed weasel is larger and has a much longer, black-tipped tail. This species is found in the Midwest, also east into Pennsylvania and West Virginia, and south through the Appalachian Mountains. It usually occurs in grassy and brushy fields, and in marshy areas, but is also found in deep forests in the southern Appalachians. It is found in the same areas as voles. It was originally thought to be a strictly New World species, and was called *Mustela rixosa*, but now is considered to be the same species as the European least weasel.

It is primarily nocturnal, but is often abroad during the daytime. Its home range is less than two acres. Least weasels run about their home range when foraging, investigating every burrow entrance or other opening. They frequently stand on their hind legs and survey their domain. Their major food is meadow voles, but they take mice or an occasional shrew, and will also eat birds eggs and some insects. The weasel chases voles through their runways, pouncing on them, and killing them with a bite to the base of the skull.

The least weasel is the smallest carnivore in North America and one of the most ferocious. It weighs only about 1/10,000 as much as the Alaskan Brown Bear, the largest North American carnivore which can approach one ton.

The least weasel dens in the abandoned burrows of other small mammals, such as

△ Note the tiny tail which distinguishes this species from other weasels.

mice or voles, using the rodent's grass nest but lining it with hair or feathers from its prey. It gives a shrill squeaking call when disturbed, and it may hiss when threatened. The female has a trill for calling the young. These animals hear best up to 16 khz, and again from 50 to 60 khz , in the range of mouse squeaks. Humans hear only up to about 20 khz. Foxes, cats, and owls are its chief predators. Like other weasels, this species will kill available prey, whether hungry or not. Such kills will be cached for later used. Caches are near the kill site or are often in side passages of the burrow.

Delayed fertilization does not occur in the least weasel which has up to three litters per year of up to six young each.

LONG-TAILED WEASEL
Mustela frenata

This is the largest weasel of North America excluding minks and ferrets. Its summer color is brown above, white below, and the tail is brown with a black tip. The short-tailed weasel is smaller with a shorter tail and white feet in summer. The mink is much larger and has a much bushier tail. Long-tailed weasels are found in a wide variety of habitats; forested, brushy, and open areas, including farms, but they like to be near water.

In the northern part of their range, long-

△ The long-tailed weasel becomes white in winter only in the northern part of its range.

tailed weasels turn white in winter, with only the tail-tip black, but further south, some molt to white while others remain brown. Considerably less than half the weasels become white in Pennsylvania, and south of about the Maryland-Pennsylvania border, none turn white. The color change is evidently genetically determined. If a northern weasel is captured and taken south, it still turns white in winter, and a southern weasel transported north remains brown. The time of the molt is governed by day length. Weasels are mottled before the color change is completed.

Females mate in their first year, before they leave the natal den, thus one male may mate with both a mother and her female babies. Four to nine young are born in early May. The maternity dens are often in abandoned dens of other small mammals. At about five weeks, eyes open, the young are approaching the color of the adults, and they are weaned. The young disperse at about seven to eight weeks, when males are already larger than their mother.

Weasels are wholly carnivorous, preying mainly on mice but also taking rabbits, chipmunks, shrews, rats, birds, and sometimes poultry. They will even eat earthworms or insects. They often hunt by zigzagging from one burrow or hiding place to the next, and may cover several miles in a night, although the actual land covered may be only 20 to 40 acres. They usually attack smaller prey, but sometimes attack prey several times their size, but this may be dangerous as they are sometimes hurt or killed. They attack by rushing the prey, wrapping their body around it, then biting it at the base of the skull. The head and thorax are eaten first, and excess food is stored. They sometimes climb a tree after a squirrel. In a hen yard, they often kill far more than they can eat at one time. They cache the remainder. They often cache several mice under a log or in a burrow. Though it seems bloodthirsty to us, this behavior makes sense, as weasels live in a situation where their next meal is not assured. They often lap blood from freshly killed prey. Weasels tend to kill whenever food is available. They then can go for fairly long periods without food if necessary.

Weasels move by a series of bounds, with back humped at each bound and with the tail horizontal or with the tip raised at a 45 degree angle from the ground. Weasels are so serpentine in their movements, that they sometimes seem to flow over rocks and logs. Their slim bodies fit through small cracks or into small burrows.

Weasels squeak and hiss when annoyed, make a purring sound when content, and sometimes utter a rapid chatter. Females may give a reedy, twittering call at mating time.

During the mating season or when alarmed, their anal glands may release a powerful, bad-smelling odor. Sometimes a weasel drags its rump, presumably to scent mark, thus informing other weasels of its sex and perhaps even its identity.

MINK

Mustela vison

The mink is long and sleek-bodied like a weasel, but it is much larger with a longer, much bushier tail and lustrous fur. It is uniformly chocolate brown to black with white spotting on the chin and throat. The marten has a longer tail and an orange throat patch; weasels have light under-parts.

Minks are found along various bodies of water: rivers, creeks, lakes, ponds, and marshes. One can sometimes see where they plunged into snow after prey, and they may make a trough in the snow, similar to an otter slide but smaller.

Mating occurs anytime from January through early April. Mating is prolonged, and occurs several times with intermittent resting. The male grasps the female by the nape during mating, but a fight may ensue if the female is not receptive. The males mate with several females but eventually stay with one. The young are blind and naked at birth and number three to six. They are born in a fur-lined nest in spring. The young remain with their mother until the family disperses in fall.

Minks of both sexes protect their territories against intruders, and males fight viciously in or out of breeding season. They maintain hunting territories by marking with a fetid dis-charge from the anal glands, which is at least as bad-smelling as a skunk's, although it does not carry as far.

Minks swim very well, often hunting in ponds and streams. They can climb trees but rarely do so.

Like weasels, minks kill by biting their victims at the base of the skull. Muskrats are often eaten although they are large and dangerous to capture, but many rabbits, mice, chipmunks, fish, snakes, frogs, young snapping turtles, and marsh-dwelling birds are taken. Minks occasionally raid poultry houses. They eat on the spot or carry prey by the neck to their dens, where any surplus is cached. Minks can hear sounds in the high frequency range of their potential rodent prey.

Minks may dig their own den in the face of a bank along a watercourse. It is about four inches in diameter, and may include underwater entrances. However, mink also den in other protected places near water, often in a muskrat burrow, an abandoned beaver den, or hollow log. All dens are temporary as mink move frequently.

Foxes, bobcats, and great horned owls prey on mink. When alarmed, mink may hiss, snarl, or screech and discharge their anal glands. They sometimes produce a purring sound when content. Pelts are highly valued. However, most of those used commercially come from minks raised on ranches.

RIVER OTTER

Lutra canadensis

The river otter is a highly aquatic carnivore. Everything about the otter is geared to an aquatic lifestyle. The feet are webbed. The otter body is streamlined with a rudder-like tail. The ears and nostrils have valves which keep water out. Otters are found primarily along rivers, ponds, and lakes in wooded areas, but sometimes roam far from water.

The river otter is active by day if not disturbed by human activity. Otters can swim rapidly. To observe its surroundings, an otter will tread water and project its head high above the surface. It can dive to about 20 feet, remain submerged for several minutes, and cover a quarter of a mile underwater. The otter can run fairly well on land, moving by loping bounds. Territories of males are larger than those of females, and may include those of one or more females. Members of the same sex are excluded from territories.

Otters feed mainly on fish, but also eat small mammals, such as mice and terrestrial invertebrates. A pair of river otters may work together to drive a school of fish into an inlet where they can be caught easily. Large catches are carried to land to be eaten, while smaller ones are consumed in the water.

Among the most playful of animals, a lone river otter may be seen rolling about, sliding, diving, or "body surfing" along on a rapid current. Otters are sociable most of the year, but during breeding season competing males may fight. Fishermen often suspect otters of depleting game fish, and otters will eat them when available. However, otters are more apt to attempt to capture slower-moving "trash fish," which are caught more easily. Because of their beautiful fur, otters were excessively trapped, leading to extirpation in many areas.

SKUNKS

EASTERN SPOTTED SKUNK
Spilogale putorius

The spotted skunk is similar to the familiar striped skunk but is smaller. Its black and white pattern is different, but still warns predators to stay away. In the area included in this book, this species occurs only in western Maryland and extreme south-central Pennsylvania, where it occurs in mixed woodlands, open areas, scrub, and farmlands.

The spotted skunk nest may be in a hollow log, under a foundation or roots, in a wood-chuck hole or in other protected places. The fur pattern of the young is already evident at birth.

Faster and more agile than the larger skunks, the spotted skunk is also a good climber, ascending trees to flee predators and occasionally to forage. Highly carnivorous, the spotted skunk feeds mainly on small mammals but also eats grubs and other insects, as well as corn, grapes, mulberries, and other plants.

Most carnivores will kill and eat spotted skunks if they can do so without being sprayed, but they usually retreat when the skunk starts its threat display. The great horned owl, its chief predator, can strike from above without warning to carry off a young skunk before its mother can spray.

THE SKUNK'S POWERFUL SCENT

When a skunk is confronted, it arches its back and stamps or shuffles its front feet as a threat. It sometimes also hisses. If that doesn't work, a spotted skunk rears up on its front legs in a handstand, its hind legs in the air, making the skunk appear much larger. If the attacker does not heed these warnings, the skunk will then release a foul-smelling stream of musk from its anal glands. Sometimes a skunk will spray while doing the handstand, but more often it squirts only after returning to all four legs, with its body twisted into a U-shape so it can see its enemy as it sprays. The skunk fires by contracting the muscles around its anal glands. Each gland has a sack which holds about a tablespoonful of a yellowish, oily liquid. Skunks only spray in self-defense. They are capable of about five or six sprays. In a few hours they can spray again. Skunks are fairly accurate up to ten feet. They aim at the attacker's face and usually hit their target. The liquid burns the eyes and may cause temporary blindness, but it has no permanent effect. The spray clings to whatever it touches, and the smell can last for weeks. Vinegar, stale beer, to-mato juice, or baking soda will help remove the skunk's odor. Washing with carbolic soap is the most common remedy. Small amounts of skunk or civet musk are used in some of the world's most expensive perfumes. These tiny amounts of musk help the perfumes hold their fragrance much longer.

STRIPED SKUNK
Mephitis mephitis

The striped skunk is familiar to most people in its range. It is a striking black and white animal, with individuals varying from mostly black to mostly white. Most striped skunks are black with two broad white stripes on their back meeting on the head and shoulders. There is a thin white stripe down the center of the face. The striped skunk is found in a variety of habitats, from woodlands and grassy plains, to suburbs and agricultural lands.

Striped skunks often den in a burrow abandoned by another animal, especially a woodchuck, although they also dig their own and may use any protected place, such as a hollow log, crevice, or beneath a building. Skunk hairs at the entrance of a burrow and small pits dug in grass or clawed-up earth are often signs that a skunk has been using a burrow or foraging for ants or grubs.

Mating is in late winter. In mid-May, four to seven young are born. They are blind, but they have very fine hair clearly marked with the black-and-white pattern. The young are weaned at six to seven weeks. At this time, the scent has developed but is not yet very potent. The procession of a mother followed by her young in single file is a site to behold.

While most mammals have evolved coloration that blends with their environment, skunks are boldly colored, a warning to potential enemies to steer clear. The smell of skunk spray may carry a mile. A sudden movement or noise, or too close an approach by an intruder, can trigger the spray, and the striped skunk can spray even when held up by the tail. A mother skunk is fiercely protective of her young, and at the approach of an intruder will snarl, stamp, raise her hind legs, click her teeth, and finally, arching her tail over her back and turning her rump toward the enemy, brace hind feet to spray.

Striped skunks are omnivorous, feeding on a wide variety of vegetable matter, insects and grubs, small mammals, eggs of ground-nesting birds, and amphibians. Also, they will feed on many kinds of fruits, blackberries, strawberries, apples and the like. The author once observed a skunk in a persimmon patch with its entire snout completely orange with persimmon. Eggs of ground nesting birds are eaten. The eggs are broken by the skunk by rolling them between their hind legs until they hit upon some hard object. Although they feed on many insect pests, they also root up lawns in the process. Mothballs sprinkled on the ground help

△ Although the proportion of white to black fur varies, the classic pattern is two white stripes that meet at the head. Above are two young striped skunks.

to discourage them from visiting homes or campsites.

Striped skunks do not hibernate, but they do fatten up in the fall prior to winter. During extremely cold weather, they may become temporarily dormant. Skunks conserve energy during the winter by spending little time above ground, sleeping a lot, and by communal denning. They may lose up to half of their weight during winter.

The striped skunk and the raccoon are currently the chief carriers of rabies in the US. They have supplanted the dog. Skunks can get either the dumb or furious form. Skunks are primarily nocturnal. If a skunk appears in the daytime, it may be rabid or have distemper.

The great horned owl is the only serious predator of the striped skunk.

Lynx capturing a showshoe hair. Lynx populations are cyclic, peaking about every nine to ten years, paralleling that of the snowshoe hare, which is 75% of its diet.

INTRODUCTION TO CATS
Family Felidae

Cats have long, sleek bodies, powerful legs, short heads with small rounded ears and eyes that face forward, providing binocular vision and depth perception. Cats cannot see in complete darkness. However, they can see quite well in the dim light of night. Their pupils contract into vertical slits during the day (protecting the eye from too much light), but expand to nearly fill the eye at night. Cats have a layer of cells behind the retina that is sensitive to even the dimmest light. Cats also use their sensitive whiskers to determine the width of spaces and whether they can pass through. Their teeth are sharp-pointed for shearing meat. Their rough tongues are used to groom their fur, and also to rasp meat from bones. The five toes on the fore foot and four on the hind foot have retractile claws. The claws are usually withdrawn, but they can be extended to slash prey or an antagonist. Soft foot pads surrounded by fur permit stealthy stalking. Cats can swim, though most do not like getting wet. However, they are excellent climbers. Most cats are nocturnal and solitary, except during the mating season. They mark out territory with feces as well as with urine and they use tree scratching posts.

LYNX   *Felis lynx*

The lynx lives in dense coniferous forest with interspersed rocky areas, bogs, swamps and thickets. In the Northeast, the lynx occurs in northern New England and extreme northern New York.

The lynx often climbs trees. By day it remains in a resting place under a ledge, among the roots of a fallen tree, or on a low branch, waiting to leap down onto passing prey. The long ear tufts serve as sensitive antennae, enhancing hearing.

The large, thickly-furred feet permit silent stalking and also increase its speed through soft snow, in which prey may flounder. Big feet also help make the lynx a powerful swimmer.

In addition to the snowshoe hair which makes up the bulk of its diet, the lynx also eats birds, meadow voles, the remains of dead moose and occasionally small, winter-weakened deer. Its diet is more diverse in summer than in winter.

△ This view of a young lynx clearly shows the solid black tip of the tail. Compare this to the tail of the bobcat on the following page. The tip of the bobcat tail is black above and white below.

△ This photo of a lynx in snow shows the extra long ear tufts which are much longer than those of the bobcat.

85

BOBCAT *Felis rufus*

Bobcats live in scrubby country with broken forest, but they adapt to swamps, farmlands, and arid lands if rocky or brushy. Rocky ledges or swamps serve as centers of activity for bobcats. Even though it is heavily populated, Massachusetts has a good population of bobcats because it has many rocky ledges.

The bobcat (and some other carnivores) marks its territory by urinating, and also uses tree trunks as scratching posts. It caches excess food somewhat haphazardly and covers it scantily with ground litter.

The bobcat has an interesting manner of walking which may be an adaptation to stalking. As it hunts, it can see where to place its forefeet noiselessly, then it brings its hind feet down on the same spots so that the prints are close together or overlapping.

Mating is in spring. The nest is of leaves or other dry vegetation and is built in a hollow log, rock shelter, under a fallen tree, or in some other protected place. There are sometimes two litters. The young remain with the mother for about a year. The bobcat is a solitary species. The only association between the sexes occurs briefly during mating. The bobcat is found only in North America, where it is the most com-

△ Note the bobbed tail with a white tip which is characteristic of bobcats. Around the world there are a number of cats with bobbed tails. They are often grouped in a genus called *Lynx*. The cat in this photo also shows the characteristic pointed, black-tipped ear tufts.

mon wildcat. It rests during the day in a rock cleft, thicket, or other hiding place. The bobcat is most active from three hours before sunset until midnight, and in the last hour before dawn to about three hours after sunrise. This animal spends less time in trees than the lynx but is an excellent climber. Like the lynx, it will rest on a boulder or low branch, and drop onto passing prey. The spotting on its fur provides excellent camouflage. It has no aversion to water and will swim when necessary. It uses the same hunting trails repeatedly, and stalks or pounces on its prey.

The main diet of the bobcat over most of its range is the cottontail rabbit, or cottontails and snowshoe hares where both occur, such as in New England. However, bobcats supplement this diet with hares, mice, squirrels, birds, reptiles, deer, housecats and many other species. They even eat skunk, porcupine and bats in caves when available. Unless prey is scarce,

they do not generally eat carrion. Small prey is consumed immediately; larger kills are cached and revisited. The bobcat occasionally preys upon livestock and especially on poultry. Both lynx and bobcat can kill full grown deer. However, most of the deer eaten are young deer, carrion, or deer caught in very deep snow.

LYNX VS BOBCAT

The bobcat is one of two short-tailed, medium-sized cats in the eastern US. The lynx is the other. The bobcat is more reddish with much more spotting than the lynx. The bobcat tail has two or three black bars, a black tip above, and is pale or white below. The lynx is grayer, the tail is black-tipped above and below. The lynx also has larger feet, longer legs, more pronounced ear tufts, and longer fur with little spotting. The legs of the bobcat have dark or black horizontal bars above. The ears are slightly tufted.

CUTE – BUT BEWARE!

Although cute and cuddly looking, even young bobcats are rarely tamed. They are ferocious animals and should not be kept in captivity. The male bobcat sometimes kills the female when they mate, an example of how vicious they can become when excited. The cute little bobcat kitten at right was brought to a wildlife rehab center because its mother had been struck by a car. It attacked everyone within range, exhibiting a tremendous capacity to slash and bite. An adult bobcat can slash through a heavy leather workboot with one swipe.

△ Notice the blue eyes of this baby bobcat. Many baby cats, including panthers, have blue eyes. The blue eyes gradually change to brown as the cat matures.

◁ An exceptional baby bobcat which has been handled from birth by an experienced animal trainer. This docile pose is very unusual and deceptive.

△ Like every other cat, bobcats have teeth which are modified for a strict meat-eating diet. Cats have fewer teeth than many other mammals, which increases the pressure exerted by each tooth. They have a shorter jaw for a more powerful bite, and all their teeth are sharp, including the molars! The enlarged canines are used to hold and stab the prey.

HORSES

Equidae

Horses are, of course, not native to the Northeast, but there is a wild population in the Assateague/Chincoteague Island area off the coast of Maryland and Virginia. The origin of these horses is not clear, but they have been there for at least three centuries, although supplemented with additional horses early in the 20th century. There are two herds at present: one on Assateague and one on Chincoteague, and each is maintained at about 150 animals. The populations on Assateague Island is regulated by birth control, while those on Chincoteague are culled by the Chincoteague Fire Department in a well-publicized roundup, and swim to the mainland.

The Assateague ponies need little description as they are virtually identical to domestic horses. Horses are very large, with a long snout, long tail, a mane and with one large, symmetrical hoof on each foot. The hoof print is large and symmetrical with a notch behind. Horses generally weigh 600 to 850 pounds; the horses on Assateague and Chincoteague may be somewhat lighter.

Horses spend about 80% of the day feeding, 20% resting, and about half of the night feeding. Horses are entirely herbivorous, grazing on green plants in summer, browsing on woody material in winter. Horses are not ru-minants, as the stomach consists of but a single compartment. There are intestinal microfauna which help to digest cellulose.

Horses form two types of social groups: territorial groups of males and females, and harem groups. Harem groups usually include one or two dominant males and five or six mares. The harem groups are very stable. Females rarely leave them, and the group stays together even if the dominant individual is lost or replaced.

The dominant individual, usually a male, leads the group to forage or to water. A dominant male has exclusive rights to the females, but must defend the harem against other males. During confrontations, the two horses stare at each other, the ears are back, they both defecate and smell the droppings of the other. Either may leave; otherwise, they move towards each other with tails high and necks arched. They sniff each other, scream, and then may fight by standing side by side. The objective is to try to kick the other off balance or until one is beaten or leaves. Young males avoid confrontation by facing dominants, but with ears up which is non-threatening.

Members of a herd practice mutual grooming by using the incisors to groom the neck, withers (the highest part of the back), and base of the mane of another horse. This helps to maintain social ties. They also groom themselves by rubbing their bodies against a tree or post, or by rolling in water, dust or mud.

The female leaves the band before giving birth. The foal is able to run with the mother shortly after birth, and mother and young return to the herd within a few hours of birth.

WAPITI OR ELK
Cervus elaphus

Wapiti is the preferred name of this animal. The use of this name avoids confusion with the European elk. Wapiti is a Shawnee word meaning pale deer, and refers to the sides of the Rocky Mountain elk which are often very pale, as in the photo above. Male wapiti weigh about 600 to 1100 pounds. Unlike most other cervids, an upper canine, the elk tooth is present.

Wapiti once ranged through most of what is now the US, but decreased due to hunting, and as agriculture replaced their habitat. The wapiti was extirpated from the eastern US in the last century. The last native animal was seen in Pennsylvania about 1867, but the wapiti was reintroduced into that state and a herd of about 250 now exists. They live mostly in mixed open and forested areas.

Wapiti are primarily nocturnal, but are particularly active at dusk and dawn. They can run at speeds up to 35 mph, but they move through the forest rapidly and almost silently. Wapiti are herding animals, especially the females and young. During the non-breeding season, cows and their young form herds, but as they approach maturity, juvenile bulls spend less and less time with the cow-dominated herds. During the rutting season (peaking in October and November), adult bulls join the herd of cows. At this time bulls give their bugling call and clash their racks of antlers in mating jousts. The jousts are ritualized and the participants are seldom injured. However, sometimes the jousts turn into actual fights and may result in injury or even death. The most polygamous deer in America, and perhaps the world, bull wapiti assemble large harems (up to 60 cows, when available).

Elk mark the areas they frequent. Seedlings are stripped of bark by cows with their lower incisors, or by bulls with the base of their antlers, and then rubbed with the sides of chin and muzzle. These posts may serve as territorial markers, warning other wapiti to keep out.

Wapiti feed on many kinds of plants but they are primarily grazers in summer, and they browse on woody vegetation in winter. Mushrooms and many lichens are also eaten.

DEER

WHITE-TAILED DEER
Odocoileus virginianus

The white-tailed deer in summer is reddish-brown above and is grayer in winter. The tail is cottony white below, giving the common name to the species. The antlers of the buck have the main beam pointing forward with up to four un-branched tines (or prongs). Antlers spread to three feet in width. Does normally lack antlers. The fawns are spotted, which helps them blend into the background when at rest. Although males can weigh up to 300 pounds, few deer weigh over 200 pounds, and most weigh much less.

White-tailed deer are found in a number of habitats, but especially in mixed woods, farmlands, and grasslands. They leave numerous signs indicating their presence, including the familiar hoof marks which are heart-shaped, and the piles of long, nearly black fecal pellets. Oval areas where they have bedded down can often be seen in grass or snow. "Buck rubs" occur when bucks rub their antlers against small trees as a territorial marker. The lower part of the trunk becomes white and polished. Deer have no upper incisors, thus do not cut twigs cleanly like rabbits, but pull them off, leaving a ragged edge.

When alarmed, the white-tailed deer raises its tail, showing the white underneath. This communicates to other deer that danger may

be present, and it helps a fawn follow its mother in flight. Whitetails snort through their noses and stamp their hooves. The snort is very commonly heard. Deer are mainly nocturnal, but may be active at any time.

They graze on green plants, including aquatic ones in the summer. Acorns, beechnuts, and other nuts, and corn are often eaten in the fall. Deer are basically grazers, but feed on woody vegetation in winter or when green plant material is not available. They browse on twigs and buds

△ Deer fawns have spotted coats which may provide camouflage for the young while they are most vulnerable.

of many different species of conifers and hardwoods.

Deer are good swimmers and can run at speeds of 35 mph. They can jump horizontally about 28 feet and vertically about eight feet. When startled, they do not run great distances but soon stop in a concealed area.

Bucks and does herd separately most

of the year, but in winter they often gather together (or yard) in areas where food is abundant. Deer yards may contain up to 100 individuals. Yards have the advantage of keeping pathways open and of providing protection from predators, but are definitely disadvantageous if the weather remains bad too long. Then the deer stay in the yards, deplete the food, and many may die.

There are two types of social groupings in deer: the doe and her young which remain together for nearly a year, and groups of young bucks. The doe group sometimes contains more than one doe, and sometimes contains two generations of young for short periods. Young males become solitary in spring and early summer of their first year, then they often form buck groups of about three to five individuals. Young females will sometimes join buck groups for short periods, and bucks move in and out of the groups, but the young bucks avoid the dominant males.

There are a number of aggressive behaviors in deer. Low-level aggression may be indicated by a direct stare along with lowered ears. Head up, head down patterns show indications to strike and chase. The chase includes kicking, and finally the dominant deer may rear and

▽ A male white-tailed deer with two females.

flail its legs at its subordinate. Deer of equal rank may rear and flail at each other.

White-tailed deer are less polygamous than other deer, and some bucks mate with only one doe. The rutting season begins after antler growth is complete and the velvet has been lost. Early in the season, males begin sparring. Antler size gives a visual indication of dominance. Sparring consists of pushing one another with the antlers with heads lowered. Occasionally, antlers become locked and the bucks may die. Courtship begins about a month or six weeks after sparring. The buck groups break up and bucks spend their time chasing does. Females usually produce one fawn the first year, However, the second or succeeding years, they usually produce two fawns, or even three when conditions are good.

White-tailed deer were once nearly extirpated in much of the Northeast and mid-west, but they are now more abundant than ever, thanks to hunting restrictions and the decline in numbers of predators, mainly wolves and mountain lions. Deer are now one of the most abundant game animals in eastern North America. Humans and the automobile, along with dogs, are their main enemies.

HOW DEER UTILIZE BARK AND LEAVES

Deer are ruminants, like cows, with a four-chambered stomach. They chew their cud, bringing food up for more chewing. Ruminants can eat roughage such as leaves or bark which is non-digestible to humans and other mammals that have simple stomachs. This fiber is then broken down by cellulose-loving bacteria which live in the stomach of the ruminant. The bacteria are passed down the digestive tract and absorbed as nutrients.

DEER

HORNS AND ANTLERS

Horns have a keratin fiber shell surrounding a bony core. Antlers are primarily of bone. Horns and antlers function as weapons and also make an animal appear larger to a potential opponent. They may also play a role in sexual attraction.

Unlike antlers, horns are not shed; they ordinarily last a lifetime (except those of the pronghorn antelope, found in the western US. The keratin outer shell of the "pronghorn" is shed annually). The bison is another example of an animal with horns. Indians and pioneers used hollowed-out bison horns to store their gunpowder.

Horn tips are often removed from domestic animals to keep them from hurting each other and to make them less dangerous to humans. If the tip of a horn is cut off, the animal will bleed from the soft tissue inside. Cutting a horn is also very painful for the animal. Ranchers now use hot irons to slice off the tips of the horns. This stops the bleeding but is still painful. Some new breeds today are "polled," meaning hornless.

In contrast to horns, antlers are shed and replaced every year. Antlers begin to grow in the spring. In the fall, they become tough and hard as they cease to grow, and the deer begin scratching them against trees and brush, removing the velvet. This is the beginning of the rut (the breeding season), when the males become more aggressive. The antlers are shed in the fall after the breeding season, and grow back again the following year. They fall off of their own weight or break off as the animal runs through brush. The break is along a separation line at the base of the antlers. antlers are seldom found in the field because they are quickly eaten by rodents.

△   Antlers showing velvet.

THE VELVET

Antlers grow from bones of the skull and during growth are covered by velvet, a soft, skin-like tissue which grows on the antlers. Velvet is rich in blood and is warm to the touch. It supplies nutrients to the growing antlers.

Only male white-tailed deer have antlers, plus perhaps one out of 100,000 females, but female antlers are not shed annually. A deer has only small antlers (tiny buttons or spikes) its first year. This first year deer is called a yearling or a spike buck. Its first full set of antlers is usually seen at 16 to 18 months. Each year the new set of antlers becomes larger until full size is reached after five to six years. Nutrition plays a large role in the size of the antlers and is even more important than genetics. In particular, the amount of calcium and phosphorus in the soil helps determine the size of the antlers. A poor diet results in antlers which are relatively smaller.

Deer antlers (along with many other deer parts) are harvested in some places for use in Chinese medicine. Whole antlers are sold in Asian medical shops. Customers slice them and boil them with chicken for use as a tonic.

THE RUT

As soon as the antlers stop growing and the velvet dries up, the deer rubs it off against trees and brush. This helps build up strength in the deer's neck muscles for jousting, and it stirs up hormones in the males that incite them to fight one another for females. This is the start of the breeding season called the rut.

Fighting among male deer becomes intense.

Sometimes two males lock antlers and cannot separate. When this happens, they may both starve to death. An area about a hundred yards wide may be completely trampled during the struggle of the two animals. Male deer in rut are so aggressive that they sometimes charge people.

SPIKE BUCK, OR SPIKEHORN

Normally, antlers drop off each year in late fall after the rut (breeding season) and are regrown again during spring and summer in time for the next year's rut. A spike buck is a yearling with its first antlers, which usually have just one point (photo below). The second-year antlers usually have two points, the third-year usually three. After three years (as in photo at left), the number of points is variable and depends on both age and nutrition.

SIKA DEER

Cervus nippon

The sika deer is native to Asia but has been widely introduced, including a population on Assateague Island off the coast of Maryland where they were introduced about 1930. They also occur on neighboring Chincoteague Island off the coast of Virginia. On Assateague Island, sika deer are found around freshwater marshes and in thickets at the edge of the woods. Sika deer are primarily nocturnal, but they often graze during the day, alone or in small herds. These deer use a stiff gallop, making bounds of about nine feet. At higher speeds they bound with all four hooves a foot off the ground at the same time. They can cover 18 feet at a bound and can clear obstacles five feet high.

Sika deer have at least ten different vocalizations. Vocalizations are used especially during the reproductive season. They include whistles between females, bleats from doe to fawn, neighs from fawn to doe and the males produce loud screams during the rut.

As is common in deer, this species feeds on green vegetation in summer, woody material in winter. On Assateague, they feed on wax myrtle, red maple, red gum, loblolly pine, largetooth aspen, pokeweed, cordgrass, poison ivy, Japanese honeysuckle, and greenbriar.

Males form territories in late summer. They dig holes with their forefeet and antlers and frequently urinate in them. The holes are about five feet wide and one foot deep and, along with thrashed ground cover around them, mark the territorial borders.

The rutting season is from September to December. Males may not feed until late in rutting. The young are born in May and June. There is usually one young per female.

△ The dark stripe down the back of the sika deer helps in identification.

93

MOOSE
Alces alces

Moose are the largest deer in the world. They are horse-sized animals with a huge pendulous muzzle. They have a large dewlap under the chin. The antler spread is four to five feet. Males weigh up to 1400 pounds.

Moose mark their presence in many ways: raggedly torn browse, thrashed shrubs, barked trees, and cleared depressions (wallows) in the ground. The wallows are about four feet in diameter and three to four inches deep. They are muddy, marked with tracks, and smell of urine. During the rut, bulls urinate in the wallow and roll in it, and cows also roll in it. Moose beds also can be seen. Their fecal pellets are more oblong than in the wapiti. Their droppings form chips or masses when they feed on aquatic plants or lush grasses. When they are feeding on woody winter browse, the pellets are pale, resembling compressed sawdust. Trails of moose are wider and deeper than those of deer, and moose are more likely to skirt obstructions or tangles of vegetation. Tracks are similar to those of wapiti but are larger and more pointed.

Moose are usually found in spruce forests, swamps, and in aspen and willow thickets. In the Northeast they have long occurred in Maine, but more recently they have extended their range through New England and presently there are even about 50 in the Adirondack region of New York.

Moose are solitary in summer, but several may gather near streams and lakes to feed on willows and aquatic vegetation, including water lily leaves. Moose nearly submerge themselves when black flies and mosquitoes bother them, or they may roll in a wallow to acquire a protective coating of mud. Moose are good swimmers, moving at speeds of six mph for up to two hours. Despite their size, they can run through the forest quietly at speeds up to 35 miles per hour. They may herd in winter, packing down snow, thus facilitating movement. They often move up and down mountains seasonally, browsing in winter on woody plants, including the twigs, buds, and bark of willow, balsam, aspen, dogwood, birch, cherry, maple and viburnum. Moose live for up to 20 years.

Vocalizations include the well known bellow of the bull and the cow's call, which ends in a coughlike *moo-agh*. Mating season is from the middle of September to late October. Bulls do not gather a harem but remain with one cow for about a week, then move to another. Bulls thrash brush with antlers, probably to mark territory. Occasionally there are fights, but generally threat displays prompt one to withdraw. Antlers occasionally interlock, some-

times causing both contestants to perish. One or two young are produced, after a gestation of eight months. The calves are light-colored but not spotted. Within a couple of weeks they can swim. At about six months, they are weaned, and just before the birth of new calves they are driven off.

A bull's antlers begin growing in April, attain full growth by August, and are shed between December and February. Moose normally avoid human contact, but they can be unpredictable and dangerous. Cows with calves are irritable and fiercely protective, and bulls in rut have been known to charge people, horses, cars, and even locomotives.

△ Antlers are among the fastest growing tissues in the animal kingdom. Within a years time they begin to grow, attain full growth, and fall off. Antlers are covered by skin (the velvet). Velvet contains blood vessels that nourish the antlers as they grow. In most species, antlers occur in males only, but they are found in both sexes in caribau. After antlers fall off, they are seldom found in the field because they are quickly eaten by rodents.

△ A mother and her calf walk along the edge of a lake at Baxter State Park, Maine.

Moose were rare in New England at the turn of the century because they were heavily hunted and their habitat was taken by man for farmland. But now because of restrictions on hunting and the decline of farming, their numbers have greatly increased. Also, in the 1970s, logging caused the return of young poplar trees along with birch and other hardwoods whose twigs provide preferred food for moose. Populations increased very quickly because moose have no natural predators besides man. It is estimated that there are now 30,000 moose in Maine, mostly in the northern part of the state. New Hampshire estimates 6,000 and Vermont 2,000.

Increased moose populations have caused some problems. Moose are hard to see at night when they wander onto highways. Headlights do not pick up their skinny legs and sometimes illuminate their elevated bodies when it is too late to avoid them. Many collisions have been reported and in most cases the result is a serious accident, some with fatalities. When a car collides with a moose, the 1,000 pound animal is usually thrown upwards, landing on the windscreen or passenger cabin.

▷ Moose in Sandy Stream Pond at Baxter State Park, Maine. In the background is Turner Mountain.

DOMESTIC CAT
Felis catus

Few domestic cats are entirely feral, meaning that they live entirely on their own in the wild without receiving food or shelter from humans. Most cats are based with humans. However, cats spend much time hunting in the wild and therefore have a great impact on small mammals and birds.

Feral and domestic cats kill large numbers of wild animals. Feeding them does not seem to reduce their killing of normal prey. Mice are their primary food. Cats also eat small rabbits and fish. Most cats do not eat numerous birds, but some cats favor birds over rodents (cooked feathers are used as a taste enhancer in some commercial cat foods). Cats are picky eaters and will actually starve with food in front of them if they do not like it.

Among mammals, cats are considered the best-designed hunters because of their stealth, ease of movement, and excellent senses. Their bodies are well designed for running down or pouncing on prey and killing it efficiently. Cats have padded feet allowing them to walk quietly, long whiskers allowing them to feel in the dark, twisting ears to pinpoint sounds, binocular vision, and great speed. Unlike the larger cats, the jaws of the domestic cat may lack the strength to crush vertebrae, but with their sharp canine teeth they can cut the spinal cord of prey without having to break the vertebrae.

Cats are color-blind like dogs, but unlike dogs, cats have excellent binocular vision. Their hearing is also very good. Cats can rotate

▲ Domestic cat stalking wildlife prey. Cats may be responsible, at least in part, for the declining numbers of certain mammal, reptile, and bird species.

their ears and they use sound to locate prey through triangulation, somewhat like binocular vision, especially at night. Cats use sound, sight, and tactile senses to hunt, but they do not rely on scent as much as dogs.

Outdoor cats, both feral and tame, are major predators of small mammals, birds, and reptiles throughout America, killing so many that they may endanger the survival of certain species. However, the wait-and-pounce hunting methods employed by cats are likely to be far more

HATE THEM, LOVE THEM: THE HISTORY OF CATS

The first mention of the domestic cat is from Egypt around 1600 B.C. In Egypt, cats were associated with religion, protected, and very highly esteemed. Mohammed's favorite animal was the cat, so cats are also held in high regard and well cared for in Moslem countries, especially since Moslems consider dogs unclean and taboo. Cats were also associated with Christianity because they were thought to have protected the Christ child in the stable from the devil.

Modern domestic cats, probably descendants of the Kaffir cat or African wild cat, reached Britain between A.D. 300 and 500. Cats later became associated with devils because of their intense eyes, and in the Middle Ages they were killed by the thousands. Superstitions such as black cats bringing bad luck arose. Cats regained popularity when the Crusaders returned, bringing rats with them, because cats became the principal means of rodent control. Still later they became a symbol of cleanliness because of their careful grooming, adding to their popularity.

successful with small rodents and reptiles than with birds. Cats have been blamed for the decline or extinction of 20 native mammals in Australia. The effects of cats on native wildlife is not fully known, however, and more studies are needed to make that determination.

DOMESTIC DOG
Canis familiaris

All dogs have their origins in wolves, from the smallest chihuahua to the huge Saint Bernard. For this reason, the behavior of dogs is largely derived from the behavior of their wolf ancestors. The dog is a social animal; it requires the company of other animals. It barks to call more dogs and often travels in packs. The bark is also a canine alarm, meant to alert other members of the pack to danger. A barking dog is calling for reinforcements ("calling out the troops"), while a dog which is ready to attack usually doesn't give any warning. Hence the saying, "Barking dogs don't bite." Packs of dogs, like wolves, have a pecking order. There is always a top dog, or alpha dog.

Feral dogs may be serious pests, especially in rural areas. Unlike stray cats, wild dogs run in packs and are threats to livestock. They are especially fond of lambs and calves. They have been known to run down deer, especially when deer get bogged down in deep snow.

EATING DOGS

In many parts of the world, especially Asia, dogs are commonly eaten. The Chinese and the Filipinos are best known for this practice. In both cultures, dog meat is considered beneficial to the health, almost medicinal. The ancient practice of eating dog meat has been embarrassing to a few modern governments. Several years ago, the National Enquirer ran a series of articles about the brutal manner in which dogs were brought to market and slaughtered in the Philippines. As a result, petitions with thousands of American signatures were *presented to then President Marcos urging a halt to the practice. Before the Seoul Olympics, the Korean government attempted to eliminate the food stalls serving dog meat in the Korean capital. In 1992, an Australian government official created an international incident by jokingly associating a friend's lost pet with the recent visit of China's premier. The joke was reported in a newspaper and the Chinese protested, regarding the incident as a racial and cultural slur. The official nearly lost his job.*

MAN

Homo sapiens

How does human behavior differ from that of other mammals? The man in the photo obviously cares for his child, but a dog or a squirrel does the same without anyone having to teach it. A mammal's urge to care for its young is instinctive. So, how much of our behavior is the result of these instincts, and how much is the result of culture, learning and experience? People have argued this point for centuries.

Certain behaviors are more influenced by instinct than others. It is also known that the higher a mammal is on the evolutionary scale, the more its behavior is likely to be influenced by learning rather than instinct. But there is always some influence from instinct, even for the most complex and advanced human behavior.

Formerly, it was thought that the ability to use tools and language clearly separated humans from the other mammals. Other traits thought to be strictly human were consciousness, learning ability, and emotion. Scientific study has shown that the gap between humans and their closest relatives, the higher primates, is not nearly as large as was once thought. Research over the past 30 years with chimpanzees and other great apes has shown in remarkable ways that they have the ability to communicate using sign language or symbols, and that they make and use tools in the wild. Furthermore, like us, they do seem to experience the emotions of pleasure, fear, rage,

sadness, depression, and anxiety.

Self-awareness, the ability to reflect on one's own condition, is generally perceived to be in the domain of human consciousness. This is a difficult trait to measure, but it is known that captive chimpanzees can recognize and examine themselves in mirrors. When asked to separate photos of animals, a captive chimp put pictures of dogs into one group and put pictures of herself in with a group of human photos. Jane Goodall, studying chimpanzees in the wild, reported that chimps are capable of intentionally deceiving other individuals, which suggests that they may be capable of self-awareness.

Aggression is often associated with animals, but no one can deny that it exists in all human cultures. In the animal world, aggression is linked with dominance, territoriality (and the resources a territory contains), and with reproductive success. The dominant males with the best territories are the ones that get the females and are able to breed.

Within human societies, aggression is controlled somewhat by rituals such as laws, customs, treaties and truces, and organized competition. It is important to remember that genetically influenced behavior such as aggression is not always something precise. Genes contain blueprints for a range of behaviors that may be expressed in a wide variety of ways, depending upon environmental influences.

Nevertheless, in spite of the similarities of human and primate behavior, great

DD/DPA

differences exist. Consider the highly developed human language and human achievements in science and art. Humans are capable of understanding the concept of events which will occur in the future and also those which have occurred in the distant past. We can understand the relationship of events. We are capable of transforming our environment to meet our needs for food and energy (although we have failed to adequately control human population growth or to preserve the environment).

MAN AND BEAST: MAN'S RELATIONSHIP WITH ANIMALS

Humans have been interacting with animals throughout their history on earth. Even today hunting societies have elaborate rituals to appease the spirits of animals they have killed or to insure that wild animals on which they depend will always be plentiful. In our own society, the issue of hunting always provokes strong reactions.

Some animals arouse strong negative sentiments. Bats are often hated animals, associated with the occult and misfortune. However, they were considered Lords of the Night by the Aztecs and are considered good fortune by the Chinese. Cougars were considered "varmints" by early settlers, but represented powerful symbols of a positive nature in the cultures of many Native American tribes.

What the role of animals in a society should be has been an open debate going back as far as the ancient Greeks. Some people felt that animals had souls while those at the other extreme believed that animals felt no pain and were no better than machines. In western society, humans

considered themselves to be higher creatures and thus justified in using animals for any purpose, regardless of the suffering of the animal. Today, most people realize that animals can experience suffering, at least those with a highly developed nervous system, and many people believe that animals are worthy of our moral concern. But in our highly industrialized world, man has lost touch with nature and animals. Perhaps this is one reason for the great popularity of zoos and the large increase in the number of pets.

The usefulness of protecting animals is being rediscovered through the conservation movement. With the destruction of much of the rain forests, it is estimated that by the end of the century, one million species will have been lost. While saving species is not of the same priority as feeding hungry people, it is important to consider the contributions of wild animals. Domestic animals have served man as storehouses of food and as work power.

The mobility provided by the camel and the horse made possible exploration of new regions. Today, many animals are important to medical research.

Domestication of animals and especially the development of farming led to the present human overpopulation. It is the factor of overpopulation, plus its associated damages to the environment (pollution, loss of habitat and biodiversity, global climatic changes, etc.) that have led mankind to the brink of global disaster, and highlighted the need for drastic action to prevent further damage to the earth. Apart from direct scientific or economic benefits, the value of preserving the natural world for its beauty and wonder has never been more clear. But animals as revered as the elephant, as rare as the tiger, as endangered as the rhinoceros, and as beloved as the dolphin, to name only a few, are still being slaughtered for their relatively minor economic value.

WHALES

BLUE WHALE
(Balaenoptera musculus)

The blue whale is the largest animal on earth. It can grow to a length of 100 feet and weigh at least 150 tons, although such dimensions are only reached in the Antarctic. In the western North Atlantic, blue whales rarely grow longer than 80 feet. For obvious reasons, few of them have been weighed. Those that have been were weighed in parts, with adjustments made for the loss of blood and other body fluids during butchering. On other occasions, scientists have made estimates based on the number of known-capacity cookers required to boil the meat, bones, and blubber of a single blue whale.

Although their numbers were badly reduced by whaling earlier this century, at least a few hundred blue whales still inhabit the western North Atlantic. They are most often observed along the north shore of Canada's Gulf of St. Lawrence and on the broad continental shelf of Nova Scotia.

△ As a blue whale breaks the surface to breathe, the tall blow is often seen first, followed by a long smooth back, then finally (several seconds after the blow) the dorsal fin and tail appear. The dorsal fin, usually less than a foot high and sometimes hardly noticeable, is set far back on the body.

INTRODUCTION TO CETACEANS

The order of mammals called Cetacea includes all the whales, dolphins, and porpoises. Among the most obvious shared features of the 80 or so species in this group are: an elongated body form, sometimes referred to as cigar-shaped; the complete loss of external hind limbs; the drastic modification of front limbs into paddle-like fins, usually referred to as flippers; and the presence of horizontal tail fins called flukes (the caudal, or rear, fins of fish are vertical rather than horizontal). The flippers are supported by shortened arm and wrist bones, while the hand and finger bones are enlongated, and the flukes are supported only by cartilage. Cetaceans are generally hairless, and their skin is rubbery. A layer of fat, called blubber, is present just beneath the skin. This blubber provides insulation for the body core and stores energy for periods when food is scarce.

There are two living suborders. The Mysticeti, or baleen whales, lack teeth. In place of teeth, the upper jaws are lined with baleen plates—fibrous structures made of modified skin tissue, rooted in the gums. The ends of these plates are frayed and interwoven, creating an efficient filtering apparatus for trapping fish and plankton inside the mouth. Baleen is what whalers used to call "whalebone."

The Odontoceti, or toothed whales, have teeth rather than baleen. These teeth vary greatly in number, form, and size among the 70 species. They can have many or few small teeth in the upper and lower jaws, or just in the lower jaws. The teeth can protrude outside the mouth as tusks—for instance, in the narwhal (Monodon monoceros) and some

of the beaked whales. One of the primary differences between dolphins and porpoises is that porpoises have spade-shaped, rather than cone-shaped teeth.

The nostrils of cetaceans are on top of the head so that the animals can exhale and inhale without breaking the swimming motion. Baleen whales have two external openings, called blowholes, while toothed whales have a single opening. In all species except the sperm whale, the blowhole or blowholes are well back from the front of the head.

Cetaceans evolved from terrestrial mammals several tens of millions of years ago. They are related to the hoofed mammals (ungulates). Most scientists agree that the two living suborders arose from the same extinct suborder—Archaeoceti.

Blue whales are uncommon visitors to the US Atlantic coast. A 66-foot blue whale washed ashore at Ocean City, Maryland, in October 1891, and it was almost 100 years before the next confirmed record of the species in US east coast waters. In 1986 and 1987, several blue whales were spotted off Massachusetts in the company of fin and humpback whales.

Unlike some of its relatives, for example the fin whale and minke whale, the blue whale has a specialized diet. It depends almost entirely on krill, small shrimp-like creatures that occur in dense swarms in productive regions of the world's oceans. One of nature's most impressive sights is a blue whale ending a pass through a patch of krill. Its grooved throat bulges taut just before the massive tongue pushes water out between the baleen plates. At such moments a whale, viewed from the air, looks a bit like an overgrown tadpole.

Blue whales are nomads. A whale seen in the St. Lawrence River one summer might be seen off Cape Cod or West Greenland the next. During the winters in between, it may visit Bermuda or the Bahamas.

The tracking of blue whales with satellite-linked radio transmitters is still in its infancy. Navy scientists, however, have used equipment designed to locate and follow submarines to study the movements of a few blue whales. This work has confirmed that blue whales cover vast distances in short periods of time.

Chris Clark, an expert on baleen whale sounds, has been working with the Navy to interpret recordings made by underwater

△ It is not unusual for blue whales to unfurl their massive tail flukes above the surface at the beginning of a deep dive. Some species of whale do not raise their flukes above the surface when diving.

Unlike fish, which have vertical tail fins and swim by moving them side to side, whales have horizontal tail flukes and swim by moving their tail up and down.

△ Although many early drawings made blue whales look fat and ungainly, modern photography has shown them to be almost slender. This aerial view of a blue whale, about to be instrumented with a radio tag by researchers in a small inflatable boat, shows the whale's long, streamlined body shape.

microphones (hydrophones). The sounds made by blue whales are at very low frequencies—so low that a human listener often "feels" them before actually "hearing" them. Here is how Chris Clark described hearing from his laboratory a whale's voice through 500 miles of ocean:

"From the very first moments there were sounds. Low, almost imperceptible sounds that resonated through the ocean like a Gregorian chant. Cathedral tapestries of music too low for my ears to perceive and so slow that I could not hear the rhythms."

FIN WHALE
(Balaenoptera physalus)

Fin whales are the longest-bodied whales regularly seen along the northeastern US. Although in the Antarctic they can be as much as 85 feet long, fin whales in the western North Atlantic are rarely longer than about 70 feet.

Fin whales are often seen in groups of three to seven animals. It is not unusual for them to feed in the vicinity of other baleen whales such as humpbacks and minkes. Although the sight of a fin whale is always exciting because of the animal's size, power, and speed, the regularity of its behavior can quickly seem monotonous. A typical sequence is for the whale to surface three times in rapid succession, each time exhaling explosively, then submerge for three or four minutes. Fin whales almost never lift their flukes clear of the surface as they dive, and they breach only very infrequently.

There are at least several thousand fin whales along the east coast of North America, and they occur across the entire continental shelf, with no obvious preference for a particular depth. They are regularly seen by whalewatchers who go out from shore sites between Maine and Maryland. Most of the fin whales that move along the US east coast probably remain in the western North Atlantic their entire lives. A few, however, are known to relocate to at least as far away as Iceland.

During the early 1980s, scientists in New England began compiling a catalogue of recognized individual fin whales. The whales are identified on the basis of unique dorsal fin shapes, pigmentation patterns, or scars. The catalogue now numbers many hundreds of individuals, and much is being learned about how long these animals stay in one area, how far they range, and whether they return to the same areas year after year. There is no doubt

△ The body of the fin whale is, like the blue whale's, slender and streamlined. While the blue whale's pigmentation is mottled, the fin whale's is solid gray on the back, with an elegant "chevron" of light gray behind the blowholes and swirling brush strokes of light gray above the flippers. This mother and calf, surfacing synchronously, exhibit the typical profile of their species.

that fin whales can travel long distances fast. One that was radio-tagged near Iceland covered 180 miles in a single day. At the other extreme, a particular fin whale was photographed on 28 days during an eight week period in the vicinity of Mt. Desert Rock, Maine, suggesting that she had settled in as a "resident" of this area for the summer of 1985.

Calving is thought to take place mainly from October to January. Births have not been observed, but newborn fin whales occasionally wash ashore dead, especially along the coast between Long Island and Cape Hatteras. Mothers closely accompanied by young calves are seen regularly.

Fin whales are not as picky as blue whales about what they eat. They are just about as likely to feed on schooling fish (like herring or capelin) as on small or large zooplankton.

MP/TDR

△ The fin whale's undersides are white. This whiteness intrudes onto the right side of the head, but not the left. One good way of helping to distinguish a fin whale from a sei whale is by observing the lower jaw coloration. If the right side is white and the left gray, the whale is probably a fin whale. If both sides are gray, it may be a sei whale.

CONSERVATION OF WHALES

"Save the whales" was a rallying cry that helped define the environmental movement in the early 1970s. The International Whaling Commission (IWC), a body created in 1946 to oversee the whaling industry and conserve whales, was the target of letter-writing campaigns and street protests. It was criticized for failing to protect either the industry or the resources. Several commercially valuable species had been driven close to extinction. To remain profitable, the industry simply lowered its standards. The sei, Bryde's, and minke whales—once considered too small to waste time and equipment hunting—replaced the larger right, blue, fin, and humpback whales as the main targets. Under the IWC's ineffectual management, even the stocks of these smaller species were rapidly declining.

By the early 1980s, the idea of a global moratorium, or ban, on commercial whaling had gained momentum. In 1982, the majority of IWC members voted in favor of it. Today, ironically, the IWC is viewed by some countries as having gone too far in the direction of protecting the whales, and in the process having neglected its responsibilities to support the whaling industry. Canada and Iceland withdrew from the Commission over this issue. Japan and Norway reluctantly continue to pay their dues, while promoting within the IWC a resumption of whaling. Norway, in fact, having lodged a formal objection to the moratorium decision, allows its whalers to continue hunting minke whales in the eastern North Atlantic. As of 1997, Norway stood alone as the only country in the world with an active commercial whaling fleet. (Japanese whalers kill several hundred minke whales in the Antarctic and western North Pacific each year as part of a scientific research program.)

With the decline of whaling, some stocks have shown encouraging signs of recovery. In the North Atlantic, humpbacks are the best example. The population that migrates along the US east coast has been increasing at a rate of more than 5% per year, and now numbers about 10,000.

At the same time, it has become evident that whale conservation requires more than just putting a stop to excessive deliberate killing. Right whales offer the clearest example. Even though they have been fully protected from hunting for almost half a century, the population remains small (only about 325). Entanglement in fishing gear and accidental collisions with ships seem to be causing enough mortality to offset reproduction. As a result, the right whale population in the western North Atlantic is either growing only very little or not at all. It may even be declining.

Conservationists have come to recognize that whale populations are affected by many kinds of human activity besides whaling. Global warming, pollution, marine debris, underwater noise, military exercises, ship strikes, and overfishing are among the factors that could inhibit the recovery of depleted whale and other marine mammal populations. Only by doing a better job of regulating our own species' growth, conduct, and consumption patterns can we hope to restore, for the benefit of our children and grandchildren, some semblance of the abundance and diversity of marine life that our own grandparents knew.

SEI WHALE
(Balaenoptera borealis)

Although the scientific name of this species, *borealis*, means "northern," the sei whale (pronounced "say") has a global distribution. Interestingly, in the northern hemisphere it moves less far north than several of its near relatives – the blue, fin, and minke whales. In tropical waters, Bryde's whale (*Balaenoptera edeni*) is far more common than the sei whale. Since the two species are difficult to tell apart, many references to sei whales in low latitudes are misidentified Bryde's whales.

In the western North Atlantic, sei whales are most abundant in deep offshore waters from the vicinity of Cape Cod northward to central Labrador and southern Greenland. Within this area, however, they are most common on the Nova Scotian shelf. They are rarely seen by whalewatchers in New England or elsewhere along the US Atlantic coast.

The movements of sei whales are notoriously unpredictable. Since they feed primarily on tiny zooplankton (particularly copepods and small euphausiids), sei whales must home in on dense concentrations. Zooplankton become concentrated in response to ocean currents, temperature, and salinity, so these factors also influence the movements of sei whales.

The baleen of sei whales has finer bristles, or fringes, than that of most other whales. The only species with similarly silky bristles is the right whale. This similarity has led scientists to wonder whether sei whales might compete with right whales for the same swarms of prey. In spring 1975, scientists from Woods Hole Oceanographic Institution in Massachusetts had an exceptional opportunity to watch the two species feed together. The scientists circled overhead in a small airplane while several right whales, a sei whale, a humpback, and about 20 fin whales "shared" a large aggregation of fish (probably herring) and zooplankton. As one might expect, the right whales and sei whale ignored the fish and focused on the plankton, while the humpback and fin whales concentrated on the fish. The right whales swam at a steady slow speed (up to three knots), mouths open, their heads sometimes breaking the surface. The sei whale swam much more erratically, changing speeds and directions, veering away from the edge of the plankton patch and then charging back into it. The sei whale would start a feeding pass with its mouth wide open, then gradually close it while still moving through the dense swarm of prey. When it looked as though the sei whale was on a collision course with a right whale, the latter simply slowed down to let its faster "competitor" go first. In this instance at least,

△ This remarkably clear view of the head of a sei whale shows the single ridge between the blowholes and the tip of the jaws. Also, note that the right lower jaw is gray—just like the left side. This symmetrical head coloration distinguishes the sei whale from the fin whale.

△ Sei whales are often "skim feeders," meaning that they swim near the surface at a shallow angle with the mouth at least partly open. In this photo, the whale is on its left side. The palate, or roof of the mouth is visible as a small, triangular, pink area in the middle right part of the image.

it appeared that there was plenty of food for all.

Unlike fin whales, which they resemble in overall appearance, sei whales do not usually arch their backs high above the surface as they dive. Rather, they seem to intend to go only a few feet or yards deep, so their back stays relatively straight.

Sei whales are about 15 feet long at birth compared to about 20 feet for fin whales. Sei whales do not grow longer than about 50 feet.

MINKE WHALE
(Balaenoptera acutorostrata)

The minke whale (pronounced "minky") is the smallest of the baleen whales in the western North Atlantic, reaching a maximum length of about 28 feet and weighing less than 10 tons. Its head is relatively short and pointed, with a prominent ridge between the blowholes and tip of the jaws. There is a sharp division between the dark gray back and sides and the white belly and throat. The most striking feature of the minke whale, and one of the best ways of identifying it in the field, is the bold white stripe on each flipper.

Minke whales are, at times, more animated in their behavior than the other members of their genus. They move quickly and change directions often, sometimes lunging or breaching above the surface. They can also show curiosity toward slow-moving or stationary boats. Whalewatchers often scan the horizon for a distant blow, only to look down and see a minke whale passing alongside the vessel, rolled onto its side as if to improve its own view of the two-legged visitors.

Because of their relatively small size and solitary habits, minke whales are hard to spot. Their blow is usually not visible, so the usual cue is just a back and dorsal fin, exposed briefly as the animal rolls into its next dive. Perhaps the best opportunities for spotting minke whales come when several of them have discovered the same school of prey and are therefore feeding in the same area.

Although they do eat plankton, minke whales are mainly fish eaters. Their habit of chasing herring can lead to serious problems, especially in "downeast" Maine and neighboring New Brunswick where weirs are set alongshore to trap schools of herring. Young whales seem especially prone to run afoul of fishing gear, and they often get no second chance to show that they've learned the lesson of avoiding such hazards.

Minke whales are abundant but scattered throughout the Gulf of Maine during spring and summer. The peak period for sightings in Cape Cod Bay and Massachusetts Bay is from July through September. Many minke whales apparently migrate offshore and farther south in winter, which is thought to be the calving and mating season. Surveys in the early 1990s suggest that at least 2000 minke whales are present in the northern Gulf of Maine and Bay of Fundy during summer.

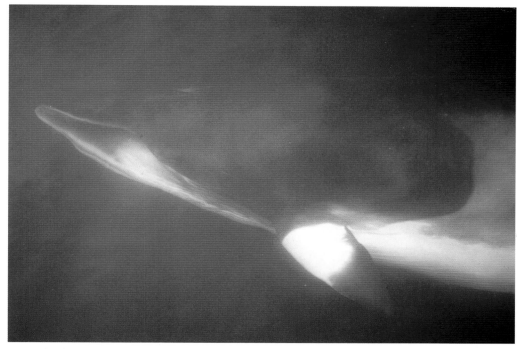

△ The broad, white stripe on the flipper readily identifies this as a minke whale. The swath of light-gray that sweeps up onto the side of the body behind the flipper is also typical of minke whales.

RORQUALS

The blue, fin, sei, minke, and humpback whales are all rorquals. The name "rorqual" comes from a Norwegian word meaning furrow. It refers to the pleats, or grooves on the chest and throat of certain types of whales. These ventral grooves make it possible for the throat to expand greatly during feeding when the whale takes in tons of water. The throat balloons dramatically and then shrinks back down to normal size as water is filtered through the baleen plates, trapping prey inside the mouth.

MONSTERS OF THE DEEP OR GENTLE GIANTS?

Many Americans formed their first impressions of whales by reading Herman Melville's classic novel, Moby Dick. This grand tale of obsession portrays a terrifying white sperm whale that bit off a whaler's limbs, carried whalers off in its jaws, and sank whaling vessels by ramming and capsizing them. In reality, at least one whaling ship, the Essex, was sunk by a whale that rammed it repeatedly. Thus it was not uncommon in times past to think of whales as monsters of the deep. Perhaps they were precisely that to the men who stalked them from sailing vessels far away from the safety of land. Today's popular image of the whale is much friendlier. Trained captive whales entertain at tourist attractions; captives have been given names and birthday parties are held for their calves. Movies and TV shows such as "Flipper" and "Free Willy" depict whales and dolphins as friends of children. Magazine photos show divers interacting with large whales, including sperm whales like the legendary Moby Dick. Because of their wholesale slaughter in the past, whales have become popular symbols of man's destructiveness toward nature. The pendulum has swung far away from fearing and dreading whales. Definitely, they are now viewed as gentle giants rather than monsters.

HUMPBACK WHALE
(Megaptera novaeangliae)

The humpback whale was unfortunate in being large, slow-swimming, and coastal in its distribution. These characteristics made it an attractive target for whalers. On the other hand, these same features, along with its black-and-white markings, elegantly long flippers, and acrobatic behavior, have made the humpback the most popular whale in the world among whale-watchers.

The scientific name means "large-finned New Englander," the first part obviously referring to the long, flexible flippers. Humpbacks are excellent subjects for research. During the early 1970s, scientists in New England discovered that each whale's tail flukes were marked in a unique manner—partly due to pigmentation and partly to scars and scratches accumulated during the animal's lifetime. At the beginning of a long dive, humpbacks lift their flukes clear of the surface, as if posing for an ID photo. Differences in the size and shape of the dorsal fin can also be useful in telling one whale from another. A catalogue of humpback whale "flukeprints" has been accumulating at the College of the Atlantic in Bar Harbor, Maine, for almost 25 years. The pictures have allowed scientists to map the movements of many hundreds of individuals and to estimate the size of the humpback population. They have also made it possible to measure, with remarkable accuracy, such elusive things as how many months a mother nurses her youngster, how long it takes for the calf to reach its own maturity, and how many years pass between births by a single female.

△ Feeding humpback whales often surge through the water surface as they capture their prey. In this photo, a group of at least four humpbacks erupt through the surface in a tight group, their huge mouths agape, allowing very little room for their prey to escape.

△ Prominent knobs line the rostrum (the area in front of the blowholes) and the lower jaws of the humpback. Whalers called them "stove bolts" as they resemble rivets arranged in rows on the whale's head. Each knob is actually a hair follicle and has one short, coarse hair growing from its center. These hairs presumably have some sensory function.

△ Occasionally, a loud report echoes across the seascape, followed shortly by more. A whale is lobtailing—pounding the surface with its powerful tail, making a mighty splash to accompany the smack of contact. Lobtailing may have a signaling function—perhaps indicating annoyance, a gesture of territorial defense, or a way of announcing "I have arrived" or "I have found food, come join in the feast."

△ The distinctive markings on the undersides of the flukes make it possible to recognize individual humpback whales.

BREACHING

When a whale leaps from the water, the maneuver is called breaching. Not all whales breach, but humpbacks do it quite often. Right, sperm, and minke whales breach fairly frequently. Blue, fin, and sei whales either never breach or do it only occasionally. The spectacle of a whale breaching usually becomes the highlight of a whale-watching tour. Photographers must respond quickly because there is no warning (at least for the first leap). Breaches can be one-time events or occur in a series. The whale may belly flop or flip onto its back. In most cases, the tail does not completely leave the water. Because breaching takes so much energy, scientists speculate that it is not done simply for fun and probably has a more serious function. One possibility is that a whale breaches to communicate a message. For example, "Here's where I am. Keep your distance" or alternately, "Come join me." Breaching seems to be contagious. When one whale does it, other individuals in the same area often start doing it also. The loud sound created when the whale smacks the surface at the end of a breach could be a threat to other males, or it could be an enticement to a potential mate. On the other hand, whales as young as a few weeks often breach. To them, breaching may be fun. They might also just be learning by imitation.

WHALE SOUNDS

The toothed cetaceans have some of nature's most sophisticated sound-production and listening systems. Most dolphin species use whistles to communicate with one another and clicks to echo-locate objects (much as ships use sonar). Sperm whales do not whistle, but use loud clicking or pulsed sounds to communicate and probably to echo-locate. The baleen whales are not known to click, but produce a rich array of other types of sounds: moans, grunts, purrs, whoops, yips, etc. Some of the sounds produced by humpback whales are songs whose function apparently relates to breeding.

WHALES

The humpback population in the western North Atlantic migrates northward in spring from its wintering grounds on shallow banks in the West Indies. It is during winter, in these warm seas where little food is available and the whales seldom eat, that calves are born and mating is accomplished. Songs are "sung" by males on the breeding grounds in the hope of attracting receptive females. Whale songs may be similar in function to bird songs.

Some young whales actually start their journey northward from the southeastern US coast, where they pass the winter without migrating to the tropics. Although some migrating humpbacks can be seen along the coasts of New Jersey and Long Island, most travel well offshore. They have five main summer destinations, ranging from as far east as Iceland and northern Norway to as far west as the Gulf of Maine and the St. Lawrence River. In summer and fall, humpbacks can be encountered almost anywhere between Maine and Massachusetts. The summer population in this region is at least a few hundred, out of a total of about 10,000 throughout the North Atlantic.

In the southern Gulf of Maine, humpbacks feed mainly on herring and sand lance. They seem able to switch from one favored prey species to another, depending on what's most available. When herring and mackerel were depleted by overfishing during the early 1970s, humpbacks largely abandoned their usual haunts in the northern Gulf of Maine and concentrated, instead, on the southwestern Gulf of Maine, where sand lance were on the increase. As the herring and mackerel stocks recovered and sand lance declined in the mid-1980s, the humpbacks began returning to their earlier distribution. By the 1990s, they were once again plentiful in areas like Jeffreys Ledge and the Northeast Peak of Georges Bank.

△ This scene of a mother and her calf would not be observed in the northeast because humpback calves are born and nursed in tropical waters. Most underwater photographs (including this one) and video footage of humpback whales underwater come from Hawaii.

△ This humpback is on its back, belly and flippers exposed. The striped appearance of the belly is caused by the ventral grooves, which end at the navel.

JONAH AND THE WHALE

The Old Testament story may not refer to a whale at all. The Hebrew word in question is tannen, *and many translations say that Jonah was swallowed by a "big fish." Whales are mammals, and not fish. However, it is not clear that this was well known to writers in Herman Melville's time, much less to Biblical writers. Modern literature has accepted, however, that the story was about a whale and this has become one of the best-known Bible stories, especially popular with children. The story goes like this: Jonah was given a mission by God to preach repentance to the inhabitants of Ninevah. Jonah did not want this mission and fled on a boat to avoid it. God caused a great storm and the sailors on the boat blamed Jonah. They threw him overboard even though he offered their captain a large sum of money. Once overboard, Jonah was swallowed by the whale. After three days and nights in the stomach of the whale, Jonah agreed to obey the Lord and the whale spit Jonah up on land. Jonah went on to become an effective preacher. This story raises the question of whether a large whale could actually swallow a man without divine intervention and also whether a whale would be likely to do so. An incident in which a whale accidentally took an inflatable boat (with crew) into its mouth and then immediately released it, indicates that it could, but would probably not be inclined to do so intentionally.* —Editors

WHALE-WATCHING

Many people are interested in observing whales at close range and join whale-watching tours to enjoy this experience. The idea of having tourists pay to "watch" whales began in the 1950s and 1960s, when naturalists in California began taking people to sea to observe migrating gray whales. Whale-watching became popular in the Northeast during the 1970s and is now a well-established and profitable industry all along the coast from Maine to Maryland.

Humpback whales are the bread-and-butter species for east coast whale-watching. This is due, in part, to their predictable near-shore occurrence from spring to fall. But humpbacks are not just available; they are also spectacular to watch. They wave their long, white flippers in the air, or smack the surface with their powerful tail. They breach more frequently than any other whale species. They also exhibit curiosity about whale-watchers. It is not unusual, on Stellwagen Bank off Boston, for a humpback to come right up to the vessel, lift its head high above the surface, and eyeball the tourists, making them wonder who's watching whom.

Some scientists and conservationists have expressed concern about harassment of whales by whale-watchers. Several studies have been done to see how whales respond to the repeated close approach of motorboats. Although whales become used to the routine activities of fishing

△ Humpbacks feeding off Provincetown thrill a boatload of whale-watchers. As long as tour operators maintain a respectful distance, whale-watching can benefit conservation by raising awareness and generating concern.

boats and ferries, the constant attentions of tourboats can interrupt resting, feeding, and social behavior. Whale-watching guidelines have been developed by government agencies, in consultation with scientists and tourboat operators, to provide a "code of conduct" for people wishing to enjoy the experience of seeing whales in their natural habitat without disturbing them.

▷ The "large wings" to which the humpback's scientific name, *Megaptera*, refers are its long, flexible flippers. When a whale rolls onto its side or back, one or both flippers often wave in the air. Whether this flipper waving serves a purpose, or is simply accidental, is unclear. One wonders, however, whether the sensations experienced by a well-insulated whale might be cool and refreshing like those experienced by a child who extends his arm out of a car window on a hot day.

BUBBLE FEEDING

One remarkable feature of humpback whale behavior is "bubble feeding." The whale forms a cloud, net, or curtain of bubbles in the water. The bubbles serve as a barrier that blocks the escape of fish or krill, allowing the whale to charge open-mouthed through a dense concentration of prey.

JDW - EARTHVIEWS

NORTHERN RIGHT WHALE
(Eubalaena glacialis)

Northern right whales are the most endangered baleen whales in the world. The population in the western North Atlantic, numbering around 325, is centered between the Nova Scotian shelf and northern Florida. These whales migrate close to shore and can sometimes be seen from land. They move north in spring and south in fall. Concentrations of right whales occur in Cape Cod Bay and in Great South Channel directly east of Cape Cod. Delaware Bay was visited regularly by right whales in colonial times, but they are only rarely sighted there today. During summer, right whales are mainly observed in the northern Gulf of Maine, in the lower Bay of Fundy, or on the Scotian shelf.

Although northern right whales are mainly black, some individuals have large white patches on the chin, throat, and belly. They also have very broad, paddle-like flippers that in many respects are more like those of killer whales than of other baleen whales. Strange as it may sound, amateur observers have been known to mistake right whales for killer whales. Sightings of whales at sea often involve just glimpses—of a back, or a flipper and fluke—as the animal rolls onto its side near the surface. A flash of black and white, along with a large flipper and a single fluke (looking like the tall dorsal fin of a killer whale), can constitute the only memory of what was there.

Female northern right whales give birth between December and March, mainly in nearshore waters of northern Florida and Georgia. Calves are between 14 and 18 feet long at birth, and stay in close contact with their 40 to 55 foot mothers during the northward spring migration. Less than 20 calves are born each year in the western North Atlantic population, which means that every single one has great importance to conservation.

Although they no longer need to worry about whalers attacking them from camps along shore, these whales face an array of other hazards. During their migration, they pass through some of the busiest shipping lanes in the world. The US and Canada sometimes conduct military training exercises and equipment trials in coastal waters. Underwater explosions pose a danger to whales, as the shock can damage their hearing and vital organs. Several right whales die in most years as a result of collisions with ships. As if the need to dodge freighters, tankers, barges, and military vessels were not enough of a challenge, right whales must also avoid fishing gear. More than half of the whales in the New England Aquarium's catalogue

△ Many photographs of right whales underwater come from the relatively clear bay waters off Peninsula Valdes, Argentina, where hundreds of southern right whales assemble each winter and spring.

have scars somewhere on the body that can be traced to encounters with nets or fishing lines. Even if an entangled whale manages to break free, it often continues to tow bits of net or line. Besides chafing the whale's skin, such towed gear creates drag and makes swimming more difficult.

The distribution of right whales between spring and fall is dictated in large part by their preferred prey, copepods. Copepods are very small animals that feed on phytoplankton and are themselves also eaten by other, larger forms of zooplankton as well as by fish, such as herring and basking sharks.

Although right whales are the slowest swimming of the large baleen whales, they are capable of impressive long-distance excursions. On one occasion, a satellite-tagged mother and her calf traveled from the lower Bay of Fundy (near the Maine-New Brunswick border) south to New Jersey and back again, covering at least 2,356 miles during a six-week period.

△ Two peculiarities in the appearance of right whales, distinguishing them from all other whales in the northeast, are their complete lack of a dorsal fin and the presence of "callosities" on their heads. Callosities are encrusted skin, and they are always heavily infested with whale lice. Depending on the color of the lice, a whale's callosities can appear orange or pinkish as well as off-white. The shape, size, and overall appearance of a whale's callosities provide a means of identifying individuals. No two whales have exactly the same callosity pattern, so high-quality photographs can be used as natural "tags." The catalogue of right whale "mugshots," maintained at the New England Aquarium in Boston, has proven to be a treasure trove of information about the species. This photo clearly shows the baleen plates which filter food. Baleen is discussed in detail in a box on page 120.

"THAR SHE BLOWS"

This cry from the crow's nest sent whalemen scrambling into their boats. The look-out had seen a tell-tale puff on the horizon. A whale had been spotted. All the larger whales—blue, fin, sei, humpback, right, gray, sperm, etc.—make a visible blow, or spout, as their blowholes break the plane of the sea surface. Especially after a long dive, the blow is explosive and conspicuous. Sometimes it smells bad. The blow is really a cloud of spray— a mixture of sea water and oily secretions from the respiratory tract. A whale's lungs are relatively small, elongated, and highly elastic. An estimated 80-90 percent of the lung air is expelled each time a whale exhales, or blows. That air is immediately replaced as the animal forcefully inhales after blowing. The blows of right whales and humpbacks are generally lower and bushier than those of blue and fin whales. From the proper angle, the right whale's blow, and sometimes the humpback's, is V-shaped, with two divergent columns of spray, one from each blowhole.

ALL ABOUT KRILL

Most of the larger whales feed on small, shrimp-like creatures called krill. Although tiny, they swarm together in huge numbers, allowing the whales that depend on them to eat their fill. Technically, the word "krill" refers to a group of crustaceans known as euphausiids, although the term is sometimes used generically to refer to many of the other small creatures consumed by baleen whales. Many euphausiids are strong swimmers and not merely passive and dependent on ocean currents. In fact, some species can outmaneuver a scuba diver. Antarctic krill are abundant in such huge numbers that humans have recently begun harvesting them with factory ships. The abundance of Antarctic krill may be due to the drastic reduction in whale populations.

△ This photo shows a blue whale swimming on its side, lunge-feeding on krill. The huge mouth opens and the pleated throat distends, taking in vast quantities of krill-rich water. The water is squeezed out through the baleen plates which act as filters to retain the krill in the whale's mouth. (See page 111 for a close-up photo of baleen.)

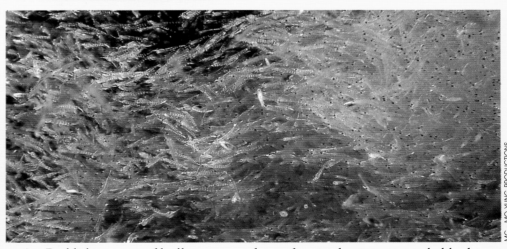

◁ △ Reddish swarms of krill seen near the surface in daytime are probably driven there by ocean currents. Such swarms can attract assemblages of predators such as birds, fish, and marine mammals.

EV/JDW

BEAKED WHALES
(Family Ziphiidae)

Beaked whales are an extremely interesting, and relatively little known, group of toothed whales. Most of the species have only one or two pairs of functional teeth, that is, teeth that erupt through the gums. In most species, tooth eruption occurs only in adult males, so females remain toothless throughout life. The males apparently use their teeth for fighting.

Six species of beaked whales are known to occur in the western North Atlantic. None of them is common on the continental shelf. Beaked whales are deep divers that feed mainly on deepwater fish and squid.

The northern bottlenose whale (*Hyperoodon ampullatus*) is the largest. Males can be 30 feet long and weigh well over five tons. Bottlenose whales live mainly in the more northern latitudes, and they are only occasionally seen off the edge of the continental shelf south of Nova Scotia. The nearest concentration area is The Gully, a deep submarine canyon not far from Sable Island, east of Halifax, Nova Scotia.

Bottlenose whales are the only beaked whales that have been the objects of a major commercial hunt. They were hunted by Norway throughout their range during the late 1800s and off Labrador as recently as the early 1970s.

Cuvier's beaked whales (*Ziphius cavirostris*) occur worldwide in tropical to temperate seas. Although not considered common off the northeastern US, they may be present with

△ Blainville's beaked whale, sometimes called the dense-beaked whale, is fairly common in some tropical or subtropical areas, but is observed only occasionally in temperate waters. In the Northeast, it is most likely to be found in the Gulf Stream or in areas warmed by this current system.

EV/HS

△ This live-stranded Cuvier's beaked whale exhibits complex facial pigmentation. Some authors refer to this species as the goose-beaked whale because of the angle at which the melon (forehead) slopes down to the short beak.

some regularity in the Gulf Stream offshore and in waters influenced by the Gulf Stream.

The other four species, all members of the genus *Mesoplodon*, are known mainly from strandings. On the few occasions when they have been observed at sea, it has been difficult to make a positive identification. It often takes an expert, who can examine the jaws and teeth directly, to distinguish among these mysterious whales. Based on stranding records, the following species of *Mesoplodon* occur at least occasionally off the northeast: Blainville's beaked whale (*M. densirostris*), True's beaked whale (*M. mirus*), Gervais' beaked whale (*M. europaeus*), and Sowerby's beaked whale (*M. mirus*).

MP/FN

SPERM WHALE
(Physeter catodon)

To many people, the sperm whale is an archetype—the most representative member of the order Cetacea. Herman Melville's famous book, Moby Dick, popularized the image of a fearsome, aggressive creature. In very recent years, thanks to the hard work of scientists, the image of sperm whales has been transformed. They are now known to be creatures of civility and intelligence.

The sperm whale is the largest toothed whale. Males grow to lengths of more than 55 feet and can weigh more than 50 tons. Females are of more modest size, growing to a maximum of about 36 feet and 24 tons. The body is dominated by the over-sized head, a feature that is especially evident in adult males. The blowhole is positioned at the front of the head, so the sperm whale's bushy blow rises at a forward angle rather than straight into the air. The lower jaw is narrow and rod-shaped. It is lined with a full battery of teeth (usually 20-26 pairs), in contrast to the upper jaw which is superficially toothless (the small vestigial teeth usually fail to erupt through the gums). Sperm whales are basically gray, often with whitish blotches on the belly and flanks. The upper lips and upper surface of the lower jaw are white.

With the possible exception of the killer whale, the sperm whale is the most widely distributed species of cetacean. It roams all oceans, from the Arctic to the Antarctic. Females and young stay mainly in tropical to warm temperate seas (where the surface water temperature is at least 60 degrees Farenheit), which means that they are less common than adult males in waters north of, say, New York. Sperm whales occur mainly seaward of the edge of the continental shelf. However, adult males in the northeast often venture into shelf waters only a few hundred feet deep.

The diet of sperm whales is dominated by squid, octopi, and deepwater fish. Adult males even consume giant squid (*Architeuthis* sp.), which are probably capable of putting up a mighty struggle before succumbing. In fact, the heads of old males are covered in white scars and scratches, at least some of which have been made by the hooks and suckers of their prey. An intact giant squid, 40 feet long, was removed from the stomach of one sperm whale at a whaling station in Australia.

Sperm whales have a complicated social structure. Calves are born after about 15 to 16 months of gestation, and they are suckled for about two years. Weaning is not as abrupt as it is for most species

△ The huge, squarish head of the sperm whale accounts for its species name, *macrocephalus*, meaning "big head." Inside the head is a large reservoir of spermaceti, a semi-liquid, waxy oil formerly prized for candle-making and illumination. It was also used to make high-grade machine lubricants. The function of the spermaceti organ in the sperm whale's head has been the subject of much academic discussion. Some have argued that it plays a role in buoyancy regulation, others say that it provides a conduit for sound transmission.

because young sperm whales grow up in close-knit communities, surrounded by cousins and aunts (all related to their mother). Adults share "babysitting" duties while mothers are away on deep dives, and lactating females apparently do not hesitate to offer milk to any youngster in the group. Males generally leave the family unit at about six years of age, joining with other males of similar age for several years of roving in so-called "bachelor schools." They become less attached to their schoolmates during their late teens and early twenties. By their late twenties, males are ready to venture into the tropics in search of mating opportunities. They move from family group to family group, usually spending no more than a few hours with any one of them. Females give birth at long intervals–four to six years for those in their prime, longer for those past it.

Sperm whales make very loud clicks underwater. During the 1950s, scientists off the coast of North Carolina thought, at first, that someone was hammering on-board their research ship. Their sound recording paper was literally blackened by the groups of 20 or so sharp clicks in rapid succession. Having established that no one on the vessel was hammering, they figured out that a group of sperm whales seen near the ship was responsible for the racket. Since those early recordings, researchers have studied the sperm whale's vocalizations in great detail. The whales exchange messages in patterned click sequences, called "codas." Long trains of clicks, often made at great depths, are probably used to locate and capture prey.

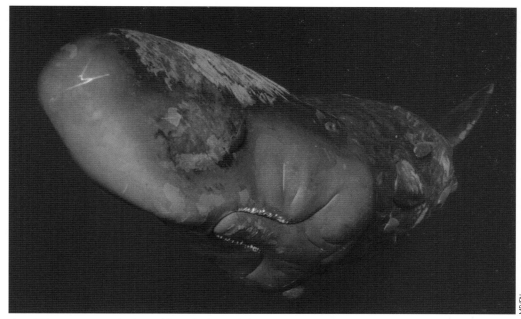

MP/FN

△ This sperm whale calf and the adult on the opposite page are "molting." This is a normal process by which whales periodically slough off old skin.

DEEPEST DIVES

Sperm whales typically dive to perhaps 1200 feet for about 40 minutes. They then spend about eight minutes at the surface resting and replenishing their oxygen supplies before diving once again. The longest "timed" dive (using underwater microphones to monitor the whale's activities) lasted well over two hours. Also, a fresh bottom-dwelling shark was found in the stomach of a sperm whale killed in an area where the water was almost two miles deep. It seems likely that the whale had to dive approximately that far to catch the shark.

KB/EARTHVIEWS

A CROOKED BLOW

The sperm whale's blowhole is positioned to the left of the centerline and at the front of the head. As a result, the blow angles forward rather than rising vertically like the blow of most other whales. In reasonably calm sea conditions, it is often possible to distinguish a sperm whale at a great distance simply because of the orientation of its blow. Such asymmetry is unusual in nature. Among toothed whales, a leftward skew of the breathing apparatus is typical, although none of the other toothed whales display it to such a degree.

AMBERGRIS: WHALE GOLD

Also called "Neptune's treasure," ambergris is a waxy material produced in the intestines of sperm whales. Most lumps of ambergris weigh no more than a pound or two, but lumps weighing several hundred pounds are found occasionally. Ambergris usually floats, so it can be found on the sea surface or washed up on shore, giving hopes of riches to beachcombers everywhere. Whalers extracted it from the colon or rectum of sperm whales that they killed. The ancient Greeks and Romans were familiar with ambergris but thought that it was discharged by underwater springs, or that it was, like honey or silk, produced by an insect. Various early peoples used ambergris as a healing drug, an aphrodisiac, a spice, and a fumigant. In the modern world, ambergris has been highly valued in the perfume industry for its ability to "fix" (intensify and stabilize) delicate fragrances. An ounce of ambergris was, at one time, worth far more than an equivalent amount of gold.

PYGMY AND DWARF SPERM WHALES
(Kogia breviceps and Kogia simus)

These two interesting species are related to the sperm whale. The pygmy sperm whale can be almost 11 feet long and weigh close to 900 pounds, while the dwarf reaches only about nine feet and 460 pounds. No one has any idea how many pygmy and dwarf sperm whales are in the western North Atlantic, but both species, especially *K. breviceps*, strand regularly along the US coast, mainly south of Cape May. On a number of occasions,

MOBY DICK

This classic novel by Herman Melville tells the story of Ahab, captain of the sailing ship, Pequod, who loses a leg in a battle with a huge, albino sperm whale. He becomes obsessed with finding the whale again and killing it. The story is told through the eyes of Ishmael, the young apprentice. Ishmael considers whales to be "spouting fish with horizontal tails," and says, "I take the old fashioned view that a whale is a fish and call upon holy Jonah to back me." Moby Dick is a treasury of information about the whaling industry in the middle of the 1800s. Melville had worked on a whaling ship and knew the subject well. However, the book did not become popular until many years after Melville's death. Poverty forced him to give up writing and take employment as a US customs inspector. Shortly after the publication of Moby Dick, petroleum products began to reduce the huge demand for sperm whale oil for lighting and other purposes, thus ending the heyday of the sperm whaling industry.

stranded individuals have been taken from the beach to an oceanarium. The animals usually lived for only a few days. These whales are difficult to detect at sea. Their blow is not visible, and their movements are slow and deliberate. They store a large amount of brown feces in their lower intestine. When startled, they almost always defecate before diving. Some scientists have suggested that the resulting large cloud of discoloration provides camouflage. Pygmy and dwarf sperm whales feed mainly on oceanic squid and cuttlefish.

KILLER WHALE
(Orcinus orca)

The appearance of killer whales, or orcas, often brings terror to marine animals, regardless of their size. In fact, other marine mammals would sometimes rather take their chances with people than face these fearsome natural predators. Seals have been known to clamber out of the water onto boats to escape, and porpoises will try to hide next to, under, or behind a vessel to keep the whales from noticing them. Even blue whales and sperm whales are vulnerable. Many large whales carry tooth-rake marks on their flukes or flippers, evidence of attacks by killer whales.

Most people are familiar with the shape and color of killer whales, thanks to having seen them in oceanaria. Their exhibition in captivity has been good "public relations" for killer whales. Rather than being hated and feared as they once were, they are now admired. In fact, some groups of animal lovers are now campaigning to have captive killer whales released back into the wild. For a number of years now, authorities, fearing public protests, have refused to allow oceanaria to collect additional killer whales from the waters of Washington and British Columbia.

Killer whales are much less common in the northeast than in the northwest. In fact, sighting killer whales is an unusual event on the US Atlantic coast. Small pods do appear fairly regularly in the southern Gulf of Maine between about mid-July and September. They are attracted partly by the schools of bluefin tuna and partly, perhaps, by the minke and other whales that arrive in the region at that time. It should be noted that not all killer whales prey on other marine mammals. In fact, different pods seem to have different specialties. Some eat mainly herring and other small schooling fish, while others rely heavily on salmon. Some pods remain in the same region year-round, making them "residents," while other pods travel long distances to match the migrations of their prey. These latter pods are called "transients."

△ Killer whales are large and robust. A breaching adult can be intimidating to a person in a small boat.

△ Killer whales live in close-knit pods, and the membership of these groups is stable over time. Hunting and foraging are usually done cooperatively—reminiscent of wolf packs and lion prides. This photo from southeastern Alaska shows some of the variability in the size and shape of killer whale dorsal fins.

JJ - EARTHVIEWS

△ The tall dorsal fin of the male killer whale may rise more than six feet in height and is the largest dorsal fin of any of the whales. In some extreme cases it may be so erect as to tilt slightly forward. Dorsal fins of females and juvenile males have a back-curved profile, similar to those of other dolphins and whales. The prominence of the adult male's dorsal fin may provide a signal to other killer whales, reminding them of his social status and perhaps discouraging them from challenging his authority.

WHALE SEX

The sight of a huge erect penis bending through the air as a group of killer whales roll against one another at the surface could mean a number of things. Young animals may simply be playing with one another or with an adult. Adult males may be engaged in homosexual activity (which may also be a dominance display). Or it may be reproductive activity between a male and female. Most cetaceans are seemingly promiscuous.

BC - MO YUNG PRODUCTIONS

ALL ABOUT BLUBBER

All whales and seals have a layer of blubber between the skin and muscle tissue. Seals gain insulation from their hair pelt, but cetaceans depend almost entirely on the blubber to insulate their body core against heat loss. The blubber performs another critical function: it stores valuable energy as fat. This function is especially significant for the migratory whales that spend part of the year foraging in food-rich areas toward the poles, and another part of the year in the tropics where they breed and give birth. Some species fast for months, and during such times they simply live off the energy stored in their blubber. Whales may get occasional opportunities to feed while on migration, and even on their wintering grounds, but their annual energy budget can only be balanced if they are

able to spend a period of time intensively feeding, often round-the-clock, on their productive northern (southern in the Southern Hemisphere) feeding grounds. Not surprisingly, the blubber is thickest at the end of the feeding season and thinnest as the animals migrate toward the feeding grounds in spring.

The thickness and the fat content of the blubber layer vary across the body. This layer is very thin or almost absent in much of the facial region and around the blowholes but a good deal thicker at mid-body. The general thickness of the blubber very much depends on the species. Bowheads can have a blanket of blubber two feet thick, whereas the blubber of dolphins and porpoises in tropical or temperate regions may be less than an inch.

Another feature of blubber is that it serves as a storage site for certain pollutants. These man-made chemical compounds enter the food chain and eventually reach a whale, or dolphin, or seal when it eats contaminated fish or other prey. The chemical make-up of these compounds causes them to bind readily with fat molecules. The best known pollutants of this kind are insecticides, such as DDT which has been widely used to reduce mosquito infestations, and the group known as PCBs, used mainly to insulate electrical transformers. High levels of DDT and PCB have been found in the blubber of cetaceans, causing concern about the possible effects on the health and reproductive fitness of the animals.

RT - EARTHVIEWS

LONG-FINNED PILOT WHALE
(Globicephala melas)

Two species of pilot whale occur in the western North Atlantic. The long-finned species is the more common of the two north of Cape May, while the short-finned species (*Globicephala macrorhynchus*) dominates in the subtropical and tropical waters south of Cape Hatteras. There are at least a few thousand long-finned pilot whales on the continental shelf off the northeastern US and thousands more to the north off Nova Scotia and Newfoundland.

Pilot whales have a specialized diet. In the western North Atlantic, they eat mainly short-finned squid and mackerel. During much of the year, these whales move along the shelf edge. However, during summer and autumn they frequently follow squid and mackerel into nearshore waters, where they are more likely to be encountered by boaters.

A young long-finned pilot whale that stranded alive in 1986 was rescued and rehabilitated at the New England Aquarium in Boston, then released off Cape Cod in June 1987. A satellite-linked radio transmitter attached to its dorsal fin provided information on movements and on the depths and durations of dives. During about three months of tracking, this whale swam an average of about 50 miles a day and exhibited an intriguing dive pattern. Long night dives were noticeably deeper than long daytime dives, although dive durations ranged from a few seconds to nine

△ Newfoundland fishermen call pilot whales "potheads" because their distinctively rounded forehead, or melon, resembles a cooking pot. This bulbous forehead, combined with their dark coloration and broad-based dorsal fin, makes pilot whales almost unmistakable

minutes overall. The whale was apparently feeding on squid at night and on schooling fish during the day. Many squid migrate vertically at night with the deep-scattering layer, a mass of organisms that responds to light conditions by moving toward the surface at night and descending to depths of more than 300 feet during the day. Apparently squid are only in shallow enough water for the whales to feed on them efficiently during the night. During daytime, it is more efficient for the whales to forage on fish that live fulltime in the surface layer.

From a birth length of about five and a half to six feet and a weight of around 165 pounds, a young pilot whale grows rapidly. Males do not stop growing until they have reached 18 feet and more than a ton and a half, at about 20 years of age. Females reach physical maturity at about 14 feet and a little over a ton, at about 13 years of age. Female pilot whales usually live longer than males—nearly 60 versus 50 years. A female gives birth to her first calf when she is about nine years old, and she generally does not give birth more

at sea. In the past, whalemen often referred to pilot whales as blackfish, and the same term was applied to several other small cetacean species as well—the false killer whale, the pygmy killer whale, and the melon-headed whale.

often than every five years. Each calf is nursed for at least three years, and the gestation period for pilot whales is about one year.

Pilot whales are among the most gregarious cetaceans. They typically occur in pods of ten or fifteen. Most pods are associated with others that, together, form loose aggregations of several hundred individuals.

Pilot whales mass strand more than any other species in the northeast (see box at right). The same pod cohesion, or tendency to stick together, that makes them susceptible to mass stranding also makes pilot whales relatively easy to hunt. In many parts of the North Atlantic, including Cape Cod during the 1800s, but now mainly in the Faroe Islands between Iceland and the British Isles, entire schools have been driven ashore by men in boats. Shouting, throwing stones, and beating the water with oars, the hunters scare the animals into shallow water, where they are easily killed with knives and lances. This operation is known as a "drive fishery."

WHALES ON THE BEACH: THE MASS STRANDING PHENOMENON

One of the most perplexing phenomena in nature is the mass stranding of whales. Scientists have long puzzled over the question: Why do they do it? It is a question that can probably never be answered completely.

All species of cetaceans strand, or wash onto the beach, at least occasionally. Stranded animals are often already dead, sick, or badly injured. Their inability to swim properly explains why they come ashore. A mass stranding is altogether different because it usually involves an entire pod or school of animals, and they are not ailing in any obvious way.

Pilot whales, false killer whales, melon-headed whales, and sperm whales are especially prone to mass stranding. The numbers involved can range as high as several hundred. Efforts by people to rescue the stranded whales are usually futile. Even if they are towed or driven back to sea, the animals too often head right back to shore. Sometimes the best option is to take some of the healthier animals to an oceanarium where they can be treated medically and rehabilitated for eventual release back into the wild.

There is no shortage of theories about why mass strandings occur. If, as some scientists have suggested, cetaceans depend on a magnetic sense to find their way in the trackless ocean, changes in the magnetic field or magnetic anomalies in particular areas could explain why the animals seem disoriented. Another possibility is that some areas function as natural "traps" due to the shape and composition of the sea floor. As creatures that

△ This photo shows a group of pilot whales stranding on a beach. Pilot whales are the most noticeable and prolific mass stranders in the Northeast. This may be related to their social structure which includes their tendency to follow a leader, even if it strands.

depend on echolocation, toothed cetaceans are vulnerable to anything that impairs their efficiency in receiving audio signals. The social bonds among members of a school usually provide benefits—for example, by helping them to avoid predators or to find and capture food. These same bonds, however, facilitate mass stranding. By following their leaders, or standing by their distressed companions, all members of a pod place themselves at risk.

△ A fin whale stranded on the shore.

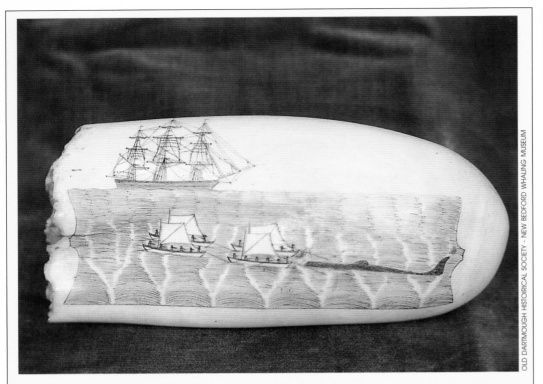

THE ART OF SCRIMSHAW

Scrimshaw, the carving and etching of whale teeth, has often been called the only truly American form of folk art. The practice originated with American whalemen who spent many months, and sometimes several years at a time, away from home living aboard sailing ships. In the sometimes long intervals between bouts of intense activity on the whaling grounds, they had plenty of time to engage in creative enterprise. The simple ingredients were readily to hand—a large ivory tooth from the mouth of a sperm whale, and a clasp knife or sharp-pointed sail needle. The raw tooth was first filed and polished to provide a smooth surface. Then the artist either scratched a design into the enamel with his knife, or created it with closely-spaced dots using a needle. The themes of the designs ranged from portraits of a wife or lover, to depictions of Bible stories, to elaborate scenes of the whale chase. To highlight his etching, the artist would rub dried pigments or ashes from the ship's fires onto the surface. This would make the illustration stand out in dark contrast to the white tooth.

The origin of the term "scrimshaw" is unknown. Early English genealogical records show that Scrimshaw was once a family name, so perhaps the craft was named after one of its first practitioners. Today, we generally refer to the whole array of objects crafted by whalers using whale products as scrimshaw—everything from clothes pins and napkin rings carved out of ivory, to walking sticks (canes) fashioned from narwhal tusks, to complicated reels for folding yarn constructed from whalebone (baleen). While purists regard the sperm whale tooth as the only authentic medium for "scrimshanding," or the making of scrimshaw, the whalers used what was available and practiced their art with walrus tusks, killer whale teeth, and even ostrich eggs. Genuine scrimshaw from the 18th and 19th centuries has acquired enormous value in private collections and museums. Large and varied displays are available for viewing at New England whaling museums, such as those in New Bedford, Nantucket, and Sharon, Massachusetts.

△ A jagging wheel for crimping pie crusts carved from a whale's tooth.

SIMPLIFYING THE NAMES

In general, everyone knows that the large cetaceans, those larger than, say, 19-20 feet, are whales. All the cetaceans with baleen instead of teeth, the mysticetes, are known as whales. The largest of the toothed cetaceans, or odontocetes, are also called whales—for example, the sperm whale, the bottlenose whale, and the killer whale. The medium-sized odontocetes are mostly known as whales, too: the pilot whales, the false killer whales, the white whale (beluga), and the narwhal (minus the "e"). More confusing are the small cetaceans, which include species called whale, dolphin, or porpoise. Strictly speaking, the name "porpoise" is reserved for those small cetaceans with spade-shaped rather than conical teeth. Fishermen in some areas refer to many additional species as porpoises, and this causes considerable confusion. True porpoises do not have prominent beaks, but then neither do many of the true dolphins. To make everything even more troublesome, some of the small cetaceans are called whales despite their small size, for example, the melon-headed whale, and the pygmy and dwarf sperm whales. These animals are all less than 15 feet in length.

BALEEN

The mysticetes, or whalebone whales, have a highly specialized way of catching their food. Tooth development stops in the embryonic stage, and instead plates, or strips, of baleen grow from the roof of the mouth. These fibrous plates of modified skin tissue are arranged in rows, one on each side of the mouth. They become frayed and interwoven at the ends, forming an efficient sieve for trapping and filtering prey from sea water. Baleen was called "whalebone" in olden times, and it was a major incentive for commercial whaling. Because of its great strength, lightness, and flexibility, the whalebone from right whales and bowheads functioned as the precursor of spring steel and plastic. It was used to make umbrellas, carriage springs, buggy whips, fishing rods, upholstery, hat frames, and brushes. Perhaps its most celebrated (and notorious) use was in the construction of ladies' wear, such as corsets, bustles, bodices, and hoop skirts. In the late 19th and early 20th centuries, after petroleum began competing with whale oil as a lubricant and illuminant, baleen became the most valuable product of Arctic whaling. In some cases, the whalers simply cut off the head and brought it on board to salvage the baleen, leaving the rest of the carcass for scavengers.

PH/MO YUNG PRODUCTIONS

WHALE-WATCHING IN THE NORTHEAST

The Northeast (here meant to include New York and the New England states, as well as the Canadian Maritime provinces of Quebec, New Brunswick, Nova Scotia, and Newfoundland) is world renowned for the whale-watching opportunities that it offers. In addition to the humpback and fin whales that are the staple tourist attractions in New England and the right whales that attract many thousands of people to the Bay of Fundy each year, the region has an array of other species that might be encountered during a lucky half-day cruise—among them minke, sei, pilot, and killer whales; Atlantic white-sided and white-beaked dolphins; and harbor porpoises. Stellwagen Bank, at the mouth of Massachusetts Bay, is something of a whale-watching Mecca. A unique feature of the whale-watching on Stellwagen is the extent to which scientific research has been blended with commercial enterprise. Scientists on board some of the vessels provide expert commentary to passengers, while taking the opportunity to observe and photograph the whales for their own ongoing studies. An estimated 400,000 paying customers go whale-watching in southern New England each year, producing revenue of about six million dollars.

The list of embarkation points for whale-watching in eastern North America is long. It includes: Montauk at the east end of Long Island, New York; Tadoussac on the north shore of the St. Lawrence River in Quebec, Canada; Grand Manan Island, St. Andrews, and Deer Island, New Brunswick, Canada; Eastport and Bar Harbor, Maine; Hampton Beach and Portsmouth, New Hampshire; Brier Island, Long Island, and Digby Neck, Nova Scotia; Trinity and St. John's, Newfoundland; and Gloucester, Plymouth, Salem, Newburyport, Boston, and Provincetown, Massachusetts. Whale-watching in the Northeast is really only

▲ A friendly humpback whale peers up at whale-watchers. Sometimes whales raise their heads above the water for a better look at the visitors, a behavior called spyhopping.

feasible between April and October. In the more northerly areas, late summer (July-September) is the best season, while a fine day in spring can be extremely rewarding off Cape Cod and Long Island. Each port has its own features, of course, some specializing in short (half-day) trips and others in longer all-day excursions. A few firms even offer money-back guarantees on whale sightings, although this is understandably rare. Part of the adventure of going to sea is that one can never be entirely certain about anything! The prospects of seeing seabirds, seals, ocean sunfish, basking sharks, and occasionally even a marine turtle add to the excitement.

THE HISTORY OF WHALING

Humans have hunted whales for more than 1,000 years. The ability to stalk, kill, and retrieve whales may have originated independently in several different locations— the Arctic, Japan, western Europe, the northwest, and possibly also the northeast coast of North America. The motive, initially, was simply to gain the means of subsistence—food for people and dogs; oil for illumination, warmth, and nourishment; and bones to serve as rafters and beams in primitive dwellings. While the smaller whales like belugas and narwhals could be taken by skilled kayak hunters working alone, the cooperative efforts of teams of whalers were needed to overcome the larger whales like bowhead, gray, humpback, and right whales. An exception was in the North Pacific, where poisoned darts were cast into a whale, with no immediate attempt made to secure the animal. The hunter simply hoped that the whale would die of its wounds within a few days, then drift to within towing range of the settlement.

The Basques of western Europe were the most enterprising early whalers. They had a well-established local industry by the 11th century and had begun sailing all the way to the New World for whales by the early 16th century. Spanish galleons were making annual visits to the shores of Newfoundland and Labrador in the 1530s, fishing for cod and hunting right whales. Dutch settlers in New York and New Jersey, and British settlers in New England, took up whaling from shore during the 17th century. They borrowed some techniques from the Basques, but they also made their own innovations. The whalemen of Nantucket are credited with being the first to pursue sperm whales offshore in sloops, and they rapidly extended their operations to distant whaling grounds. In the 1760s, American whalers figured out how to boil blubber on board their vessels, and this made it possible for them to make multi-year voyages to other oceans.

Norwegian whalers made two important inventions a century later: the deck-mounted

△ ▷ Above left, whaling ships and oil barrels in 1870. The photo above, from 1903, shows how a sperm whale carcass was cut up alongside the whaling ship. Note the long, narrow jaw, lined with teeth. The photo opposite shows a harpooned minke whale being hauled aboard a Norwegian catcher boat in the eastern North Atlantic in 1994.

cannon that fired a harpoon with an exploding grenade head into the whale; and the engine-driven catcher vessel. These made it possible to add the elusive blue, fin, and sei whales to the mix of species that could be taken. Catcher boats soon operated from processing plants on shore throughout the North Atlantic. In the first decade of the 20th century, the first floating factory appeared—a large mother-ship to process carcasses and transport whale products, accompanied by one or more fast catcher boats. The stern slipway was invented in 1926. This meant that whales could be winched aboard the factory ship for processing, rather than being cut up on shore or alongside the ship

Over the course of whaling history,

some products became obsolete and markets changed. For example, spring steel and plastic essentially replaced baleen, and fossil fuels and electricity took the place of whale oil for lighting homes and streets. But new uses were found for whale products, too. A method of hydrogenating whale oil was discovered in the 20th century, and the demand for oleomargarine provided an incentive for the massive slaughter of baleen whales in the Antarctic. Also, the heat-resistant qualities of sperm whale oil made it strategically useful as a lubricant in rockets and missiles. Today, the pressure for resumed commercial whaling comes largely from the demand for whale meat in Japan.

ATLANTIC WHITE-SIDED DOLPHIN
(Lagenorhynchus acutus)

These robust, energetic dolphins are abundant in coastal waters of the northeastern US, especially in the Gulf of Maine. Ship-board and aerial surveys have shown that there are at least 10,000, and possibly several tens of thousands, in shelf waters between North Carolina and Nova Scotia. The movements of these dolphins are not particularly well understood. They certainly enter the Gulf of St Lawrence in large numbers during summer, but those that do so must leave before winter to avoid becoming trapped in ice. It is likely that the population as a whole shifts southward, and perhaps offshore, during the colder months. Large concentrations have been reported in spring in the southwestern Gulf of Maine and along the continental slope off Cape May.

These dolphins grow to a length of about nine feet and can weigh more than 500 pounds. They are best identified by the conspicuous patch of white on the flank and the yellow or tan stripe immediately behind it.

Large schools of several hundred dolphins are sometimes observed. Such schools tend to consist of numerous smaller groups of four to fifteen closely-associated individuals. When feeding or agitated for some other reason, a herd of white-sided dolphins can put on a display of aerial acrobatics—leaping high above the surface, slapping the water with their flukes, and playing in a boat's bow wave.

Atlantic white-sided dolphins eat small schooling fish like herring, silver hake, smelt, and sand lance. The dolphins are often found in the close company of feeding fin and humpback whales—possibly preying on the same fish schools. Occasionally, a few of the dolphins will swim immediately ahead of a whale as though seeking to catch a brief "ride" in the larger animal's "bow wave."

Much of what is known about the life history of this species was learned as a result of a tragedy. In September 1974, a herd of about 150 white-sided dolphins misjudged the depth and tidal state in Cobscook Bay, Maine. The dolphins probably had been chasing a school of herring. As the water drained rapidly away from the mudbanks, the animals were left struggling to escape. In spite of valiant efforts by local people to rescue them, many of the dolphins died. The 57 carcasses recovered from this stranding were examined in detail by scientists.

Female Atlantic white-sided dolphins give birth for the first time at about seven or eight years of age. The gestation period is probably 10-12 months, and most calves are born in June or July. The birth length is about

three and one-half feet. Mothers nurse their calves for about 18 months. Females produce single calves at two- or three-year intervals. Both males and females live for 20-30 years.

△ Atlantic white-sided dolphins are aerial acrobats, at least when they are in the mood for displays like those shown above. Their bold and sharply-defined side pigmentation, particularly in combination with their stubby beak, make these dolphins easy to identify.

WHITE-BEAKED DOLPHIN
(Lagenorhynchus albirostris)

This is the northernmost dolphin. It is abundant off the coasts of Labrador, Greenland, Iceland, and Norway. Although much less common in New England than its close relative, the Atlantic white-sided dolphin, the white-beaked dolphin is seen from time to time along the US coast from Massachusetts northward, mainly in spring and early summer. Judging by accounts in the literature from the 1950s and 1960s, it appears that white-beaked dolphins have declined in abundance in the Gulf of Maine. Their decline seems to have been "offset" by a corresponding increase in Atlantic white-sided dolphins.

The two North Atlantic species of *Lagenorhynchus* are similar in size and general appearance. Both can be exuberantly acrobatic at the surface. One of the best ways to tell them apart at sea is by noting the color of the back. On white-beaked dolphins, the white areas on either side behind the dorsal fin extend onto the back, while on the white-sided dolphin the white side patches are sharply bordered and well down on the flanks. Another obvious difference is the white beak of the white-beaked dolphin.

White-beaked dolphins eat many kinds of fish (including herring, cod, and capelin), squid, and octopus. They often travel in close-knit pods of a few to perhaps 30-50 individuals. Large groups of hundreds are sometimes seen in the northeastern Atlantic.

During a cruise off southeastern Greenland, Norwegian scientists observed a close association between white-beaked dolphins and fin whales. Wherever fin whales were plentiful, large numbers of white-beaked dolphins were also present. As a whale surfaced through a ball of capelin, the dolphins appeared to snatch fish spilling from the whale's gigantic maw.

ICE ENTRAPMENT: A RARE BUT DEADLY HAZARD

The lives of dolphins in the wild are full of hazards, some natural (predators, stranding, starvation) and some man-made (gillnets, underwater explosives, fast-moving vessels). One natural hazard faced by cold-water dolphin species, like the white-beaked dolphin, is ice entrapment.

Ice entrapment of white-beaked dolphins occurs with surprising regularity along the Avalon and Burin peninsulas in southeastern Newfoundland. During late winter (usually late February, March, or early April), strong winds can suddenly push pack ice far into a bay. Dolphins that are in that bay (occasionally also blue whales or fin whales) run the risk of being crushed between ice floes or driven onto the beach. During the period from 1979 to 1991, scientists documented 21 ice-entrapment events involving a total of 320 to 350 white-beaked dolphins along the south coast of Newfoundland. More than half of the animals are known to have died. Most had severe scrapes and cuts on their bodies from sharp edges of the ice. Some had internal damage, broken ribs, and bloody froth in their lungs.

On a few occasions, attempts have been made to rescue ice-entrapped dolphins. In one case, six female white-beaked dolphins were flown from Newfoundland to a marine aquarium in Mystic, Connecticut. Unfortunately, all six died within a few months.

SHORT-BEAKED COMMON DOLPHIN
(Delphinus delphis)

Until recently, only one species of common dolphin was recognized throughout the world. Now it is clear that there are at least two species: the short-beaked form, *D. delphis*, and a long-beaked form, *D. capensis*. Only the short-beaked species occurs off the northeastern US.

The beak is "short" only when compared with that of the long-beaked common dolphin. It is proportionately long and slender compared with those of the Atlantic white-sided, white-beaked, and bottlenose dolphins.

Common dolphins are especially abundant in a broad band running parallel to the continental slope (roughly the 300-600 foot depth contour) from Virginia to Massachusetts. Schools of more than 3,000 animals can be seen on Georges Bank, directly east of Cape Cod, during the fall. Scientists comparing the distributions of common dolphins and white-sided dolphins concluded that both species prefer areas with high sea-floor relief where water depth changes abruptly. This preference is probably explained by the fact that prey organisms, like schooling fish and squid, are concentrated in such areas. The two dolphin species differ, however, in the types of water conditions that they prefer. Common dolphins are usually in warmer, more saline water than white-sided dolphins. Water masses with particular temperature and salinity characteristics

△ The short-beaked common dolphin is sometimes called the saddleback dolphin, referring to the bold, dark, V-shaped "saddle" pattern on the back, pointing down onto the sides. Viewed directly from the

△ By contrast, a long-beaked common dolphin mother with her calf. This species is not found off the Atlantic coast

move seasonally, as do the organisms associated with them. Dolphins, like most predators, must follow their prey.

The diet of short-beaked common dolphins consists mainly of fish and squid that belong to the deep-scattering layer. The activities of several common dolphins were monitored via radio-tags off southern California during the early 1970s. A regular pattern of deep diving, to depths of about 150 feet, began at dusk, just as sonar traces and tows with trawl nets showed that the

side, the pigmentation can also be seen as having an hourglass or crisscross pattern. The front half of the "hourglass" is tan or yellowish tan.

of North America. It does occur in the near-shore topical waters of eastern South America and West Africa.

deep-scattering layer was on its way to the surface. The dolphins abruptly changed their activity at first light, making only shallow dives at the same time as the deep-scattering layer began its descent.

There are many thousands, possibly a few tens of thousands, of common dolphins in the western North Atlantic. They are among the most frolicsome dolphins, often seen in schools of hundreds, whipping the sea surface into a white scar on the horizon.

STRIPED DOLPHIN
(Stenella coeruleoalba)

To an untrained eye, it is easy to mistake striped dolphins for short-beaked common dolphins. They are the same size (maximum length about eight and a half feet). Both are dark on the back and white on the belly, with fairly complicated stripe patterns on the face and sides. Their dark beaks are longer and slimmer than those of white-beaked, Atlantic white-sided, and bottlenose dolphins. At sea, the only way to distinguish striped dolphins from common dolphins is by paying close attention to the orientation of the stripes and the "cape" pattern on the sides below the dorsal fin. The striped dolphin has a bold dark stripe from the eye to the anus as well as a prominent "shoulder blaze"—a narrow brush stroke of light gray sweeping from behind the eye back and up toward the dorsal fin.

The behavior of striped dolphins at the surface is similar to that of common dolphins. Striped dolphins travel in large schools, often numbering in the hundreds. They arc gracefully through the air and whip the sea into a froth. They are deep-water animals, usually found at or seaward of the edge of the continental shelf. Like offshore bottlenose and short-beaked common dolphins, striped dolphins are most likely to be found in waters influenced by warm currents. Individuals strand occasionally in New York and New England, but most sightings of groups are well away from the coast on the outer edge of the shelf or on the continental slope.

Population estimates for this species in the western North Atlantic are well above 10,000. The striped dolphin occurs in all oceans. It has not been hunted on a large scale in the North Atlantic, but large numbers (many thousands per year) were traditionally killed in "drives" off Japan. (Drive fishing means rounding up the dolphins and herding them toward the shore, where they are slaughtered in shallow water). In recent years, Japanese

△ The complicated stripe pattern on the face and sides explains the common name of this dolphin species. Although the genus *Stenella* is fairly diverse, encompassing the spinner and spotted dolphins as well as the striped species, this is the only member of the group that regularly occurs in waters off the northeastern US.

catches have declined sharply, indicating that the kill was too large for the dolphin population to replace its losses year after year.

A FREE RIDE

Many species of dolphins "ride" the bows of vessels. The early whalers often took advantage of this behavior, harpooning a dolphin to put fresh red meat on the dinner table. Most mariners today simply enjoy watching these otherwise-independent creatures come to their vessel for a free ride. The phenomenon of bow-riding, or assisted locomotion (dolphins also catch rides in the surf and in the wakes of boats), even extends to the relationship between dolphins and whales. More than occasionally, dolphins are seen playing in the surge of water created as a large whale comes to the surface—a primitive form of bow-riding. In this photograph, a common dolphin swims just ahead of a surfacing blue whale.

BOTTLENOSE DOLPHIN
(Tursiops truncatus)

While the white-beaked dolphin lives mainly in waters influenced by cold currents, the bottlenose dolphin is more common in temperate areas influenced by the warm Gulf Stream. Bottlenose dolphins are often seen from the shore south of Cape Hatteras, but such opportunities are much less frequent along the beaches of New Jersey, Delaware, and Maryland.

Offshore, along the edge of the continental shelf in waters 600-6000 feet deep, bottlenose dolphins are abundant all the way north to Georges Bank east of Cape Cod. Small pods of these offshore dolphins are often closely associated with schools of long-finned pilot whales. Whether this relationship is related to foraging efficiency, predator avoidance and defense, or just a matter of wanting company in the ocean expanses is not known.

Bottlenose dolphins eat a large variety of organisms, probably depending on what's available to them at a given time—fish of many species, squid, and shrimp. The coastal dolphins from Cape Hatteras northward show a preference for weakfish or sea trout, Atlantic croaker, spot, and silver perch. Offshore bottlenose dolphins feed on deepwater fish and squid, at least some of which are part of the deep-scattering layer. This is a mass of organisms that responds to light conditions by rising at night from a depth greater than 300 feet.

The coastal population definitely migrates seasonally, and New Jersey is the northern limit of its range. Offshore, dolphins can be found during all seasons, including winter, along the edge of Georges Bank. It is not known whether they migrate farther north in summer.

DOLPHINS IN HISTORY

The name dolphin comes from the Greek word "delphys" meaning womb, a word origin which shows reverence for the life-giving sea. Greek gods such as Apollo sometimes took the form of dolphins. The famous Temple of Delphi and the ancient Oracle of Delphi honor Apollo in his dolphin form. Images of dolphins on coins and jewelry were often used to represent safe journeys and good luck. The fish symbol which was used long ago to mark the secret meeting places of persecuted Christians may have been derived from the religious symbolism of the dolphin. —Editors

THE DARK SIDE OF DOLPHINS

Despite their legendary reputation for friendliness, bottlenose dolphins do have a dark side. They do not hesitate to crowd smaller animals (like common dolphins) away from the most favorable positions in a vessel's bow wave. Human swimmers interacting with "sociable" wild bottlenose dolphins have occasionally experienced the less pleasant side of these animals' personalities, getting butted in the chest or pinned against the seabed. Recently, off the coast of Scotland, researchers observed (and photographed) a bottlenose dolphin tossing a harbor porpoise high into the air. The deaths of this and numerous other porpoises in the same area were traced to the actions of an exceptionally aggressive local pod of bottlenose dolphins.

PORPOISE FISHING

It is easily forgotten that in the 19th century, residents of coastal towns took pleasure at the sight of dolphins along the beach for an entirely different reason than they do today. A "porpoise fishery" operated at Cape May, New Jersey, for at least two years during the 1880s (the term "porpoise" was applied to bottlenose dolphins at the time). When bottlenose dolphins were sighted near shore, men in a steamboat would haul a large net into their path and attempt to trap them. Not much is known about the catch, except that as of August 1884, more than 200 dolphins had already been hauled ashore that season. A much larger and longer-lasting "porpoise fishery" was established along the Outer Banks of North Carolina, probably exploiting the same population of dolphins that migrated to New Jersey in summer.

The fishermen hoped to get several gallons of oil from a dolphin's blubber, plus 4-8 ounces of oil from the jaws (an excellent lubricant), altogether worth a few dollars. They also collected the hides, which were tanned and made into shoe leather. A raw half-hide (each dolphin's skin was split into two halves) was worth about $2, while the tanned product sold for $10-12. The rest of the carcass, processed as fertilizer, was worth about $2.50.

MWN

DANGERS TO DOLPHINS

Most people in North America have noticed the markings on certain cans of tuna indicating that they are "dolphin-safe." Tuna fishing in the eastern tropical Pacific Ocean was responsible for the deaths of millions of dolphins, mainly spotted and spinner dolphins but also including many other species, from the late 1950s to the early 1990s. In recent years, fishing methods have been better regulated, and the fishing industry itself has made a greater effort to avoid killing dolphins in the process of purse-seining for tuna. A consumer-led initiative to label cans was one of the strategies used to reduce the scale of dolphin mortality in the fishery. In theory at least, shoppers can expect when they buy a product marked "dolphin-safe" or "dolphin-friendly" that no dolphins were killed accidentally in the process of capturing the tuna. It should be remembered, however, that the enormous purse seine nets can never be made truly "safe" or "friendly" to the communities of sea life that they exploit. Turtles, sharks, and many other kinds of fish remain vulnerable to capture in the nets along with the tuna.

Although the so-called "tuna-dolphin problem" dominated the US marine mammal conservation agenda during much of the 1970s and 1980s, several other types of human activity are now recognized as equally serious. Here are a few examples:

Drift gillnets, some of them miles long, have been used in many parts of the world to catch salmon, tuna, billfish, and squid. These nets have been described as "curtains of death." They are usually set, or placed in the water, late in the day and hauled, or retrieved, early the next morning. Large numbers of dolphins, porpoises, seals, and seabirds are killed, unintentionally, in drift gillnets. The use of large drift nets on the high seas was banned by a United Nations resolution that took effect on January 1, 1993. However, they are still used in coastal waters under national jurisdiction. In 1998, the European Union took important steps to limit drift net fishing in the Mediterranean Sea and elsewhere in the North Atlantic.

A particularly difficult problem for the United States and Canada has been the high mortality of harbor porpoises in gillnets set primarily to catch herring in the Gulf of Maine and lower Bay of Fundy. In some countries, such as Peru and Sri Lanka, the frequent capture of cetaceans in gillnets has created markets for the meat and blubber. As a result, fishermen now go to sea with harpoons, and they also set their nets deliberately to catch dolphins. In southern South America, crab fishermen found that the meat of dolphins and porpoises made good bait, and for a number of

years during the 1970s and 1980s, the small cetaceans in southern Chile were hunted to supply this crab-bait market. River dolphins in India and Bangladesh are in demand among fishermen who use their oil as bait for a certain species of catfish. Fisheries directed at dolphins and other small cetaceans continue to take many different species off the coast of Japan, including striped, spotted, bottlenose, and Risso's dolphins. The Japanese hunts involve catcher boats with mounted guns on board (taking mainly Baird's beaked whales and short-finned pilot whales), hand harpoons thrown from the bows of fishing boats (taking mainly Dall's porpoises), and the "driving" of whole schools of dolphins and small whales ashore for slaughter. Elsewhere in the world, large numbers of long-finned pilot whales and Atlantic white-sided dolphins are killed in the "drive" fishery in the Faroe Islands; fishermen from certain islands in the eastern Caribbean Sea (West Indies) regularly harpoon pilot whales and dolphins for food and oil; and Greenlandic fishermen shoot many harbor porpoises and some dolphins to supplement their catches of fish, seals, and whales.

EV/RLP

RISSO'S DOLPHIN
(Grampus griseus)

The appearance of Risso's dolphins is almost comical, and, at times, their behavior can be downright clownish. They are not often seen off the northeast, however, except in the deep waters seaward of the continental shelf, particularly those influenced by the warm Gulf Stream. There are at least 10,000 Risso's dolphins in the population that inhabits the shelf-edge region between Cape Hatteras and Georges Bank.

The head is squarish in profile, with no beak. Functional teeth are present only in the lower jaws, which have, at most, seven pairs toward the front. The front part of the head has a deep vertical crease or furrow. The flippers are long and pointed, the dorsal fin tall and back-curved. These large dolphins can be 12-13 feet long and weigh almost half a ton.

The most striking feature is the color pattern. Basically, the body is gray, with an anchor-shaped white pattern on the chest and belly. Much of the rest of the body can be white as well, except for the flippers and dorsal fin which are always relatively dark. Adults are more or less completely covered with what seems to be a random collection of scratches and scars. Some of these are probably made by prey and some by the teeth of fellow Risso's dolphins. Like other species with reduced dentition (for example, beaked whales and sperm whales), Risso's dolphins are squid specialists.

Risso's dolphins are usually encountered in small groups of one or several dozen individuals although schools of hundreds have been described. When not busy feeding, they can

M.P./FN

△ The almost comical appearance of Risso's dolphins is matched by their antics at the sea surface. Calves are an even chocolate brown, but as they age, the skin lightens and scars accumulate. The tall, back-curved dorsal fin readily distinguishes Risso's dolphin from the beluga, or white whale. Belugas, which have no dorsal fin, occasionally wander southward from their normal subarctic and arctic range, so sightings of "white dolphins" need to be evaluated critically to decide whether they are belugas or Risso's dolphins.

be playful at the surface, coming part way out of the water and falling onto their sides, slapping the surface with their flippers and flukes, even "cartwheeling."

HARBOR PORPOISE
(Phocoena phocoena)

This small porpoise is the most common cetacean in coastal waters of the northeast. Its gray back and small triangular dorsal fin pop into and out of view so fast, however, that only attentive watchers are likely to notice it.

Harbor porpoises grow to only about five and a half to six and a half feet in length and 150–200 pounds in weight. Females are somewhat larger than males. Calves, born after an 11 month gestation period, are about two and a half feet long and weigh only 10–20 pounds at birth. They are nursed for almost a year.

Harbor porpoises do not live very long. Most die before the age of 14. Females reach maturity at about three years of age. Thus, even if they gave birth annually (which they can do for at least a few years running), they could not produce more than about 10 calves in a lifetime. Most probably produce only four or five.

The population centered in the Gulf of Maine and Bay of Fundy numbers at least 40,000. Harbor porpoises undertake seasonal migrations, but the exact routing for the animals off the northeast is not known. Winter strandings between New York and North Carolina (occasional strandings occur as far south as Florida) suggest that some of the animals move well south of their summer range, which is mainly north of Long Island. Some porpoises probably move offshore for the winter and return to the inshore bays and harbors only in late spring and summer. Satellite-linked radio-tracking has shown that individual porpoises occupy home ranges of tens of thousands of square kilometers, and that they move back and forth across the US-Canada border. One animal that was tracked from August to March initially moved southwestward out of the Bay of Fundy, following the 300-foot contour. It moved no farther south than Cape Cod, but it traveled widely within the Gulf of Maine.

Several other porpoises trapped accidentally in herring weirs have been outfitted with small time-depth sensors fastened onto their dorsal fins. The instruments, designed to fall off after about four weeks, provided information on depths and durations of dives. Dives to depths of more than 740 feet were recorded, and the longest submergence was for about five and a half minutes. A normal period at the surface would involve two to six rolls, separated by less than ten seconds below the surface. The animal would then remain out of sight for one and a half or two minutes.

The social habits of harbor porpoises are not well known. They do not generally travel in large groups, although concentrations of a few tens of animals sometimes develop in food-rich areas. Groups of three to ten are more typical of the species.

Fishermen in Newfoundland coined the term "herring hog" for the harbor porpoise. Herring are, in fact, the most important prey species for harbor porpoises in the northeast. Harbor porpoises are themselves preyed upon by great white sharks and killer whales.

△ Many hundreds of harbor porpoises die each year in the nets of American and Canadian fishermen. This is one of the lucky few that are rescued by researchers, who often use the opportunity to attach a radio transmitter before releasing the porpoise back into the wild.

MP/FN

INTRODUCTION TO PINNIPEDS

The term "pinniped" means "fin-footed." It has been applied to three families of the mammalian order Carnivora: the true or earless seals (Phocidae), the eared fur seals and sea lions (Otariidae), and the walrus (Odobenidae). Unlike the cetaceans, which are fully aquatic, the pinnipeds are amphibious, meaning that they are adapted to live both in and out of water. All pinnipeds spend at least a part of their life "hauled out" on land or ice. The term "haul out" is used often regarding pinnipeds. It can refer to the act of climbing, clambering, or rocketing out of the water and onto a platform. Or it can mean the site where the animals come ashore. Pinnipeds haul out to rest, molt (shedding and replacing hair), nurse their pups, and, for some species, mate.

The ancestors of the pinnipeds were bear-, dog-, or otter-like and apparently began adapting to life in the sea about 30 million years ago. Although most scientists today believe that all pinnipeds derived from a common ancestor, there is some disagreement on this point. Pinnipeds have front and hind limbs, modified into fins or flippers. They have nails on both the fore and hind flippers. In all, there are about 34 species of pinnipeds worldwide. Only a few of these, all phocids except for the walrus, inhabit the western North Atlantic Ocean.

HARBOR SEAL
(Phoca vitulina)

The harbor seal is, by far, the most common pinniped in the northeastern US. It is a resident of islands and inshore coastal waters from the south shore of Long Island northward. Although individual harbor seals turn up occasionally along the entire east coast, there are no regularly occupied haul-out sites in New Jersey or farther south.

Since the Marine Mammal Protection Act of 1972 came into force, harbor seals have flourished in U.S. waters. In earlier years, bounties had been paid on harbor (and gray) seals, in the belief that they competed with fishermen and damaged fishing gear. With protection from hunting, the number of harbor seals in New England has increased by a factor of about five. Close to 30,000 were counted along the coast of Maine in the early 1990s. Many more are present in eastern Canadian waters.

Groups of harbor seals are most often seen on intertidal ledges, rocky islets, reefs, mud flats, piers, and remote sand or cobble beaches. When in the water, their dark round heads bob conspicuously above the surface. They are vigilant and curious. While on land, they are usually alert and ready to slip into the water as

soon as a boat, pedestrian, or dog approaches too closely.

The daily activity cycle of harbor seals is influenced by many factors, including tidal and weather conditions, season, time of day, and amount of disturbance. They are most likely to be out of the water when the tide is low and the weather is sunny and calm. Seals especially like to bask out of the water during the molting season, when they are missing the insulation normally provided by their pelage.

Pups are born in May or June. Normally, the lanugo, or fetal coat of soft white hair, has already been shed in the mother's uterus and replaced by a darker spotted coat of short, stiff hairs. The newborn harbor seal is more advanced in its development than most baby seals. Within five minutes of birth, it is capable of following its mother out of the intertidal zone, either moving farther up the beach

or into the water, ready to swim.

The pup grows rapidly. While it may weigh only about 20 pounds at birth, the pup reaches about 60 pounds by the time it is weaned at about four weeks of age. Even though the mother seal forages during the lactation period, she loses about a third of her weight by the time her nursing responsibilities end. Mating occurs soon after the pups are weaned, and many females give birth in successive years. Harbor seals can live for 30 years or longer, although few attain ages greater than about 25 years.

Adult harbor seals are capable of diving to depths of at least 1500 feet and staying submerged for nearly half an hour. Most dives are much shallower and shorter, however. They have a varied diet, including many kinds of fish, crustaceans, squid, and octopus.

GRAY SEAL
(Halichoerus grypus)

Gray seals can be twice as large as harbor seals. Adult males can weigh more than 750 pounds, females more than 425. The two species are often found in the same areas. Apart from the size difference, which of course is not so obvious when comparing young gray seals with adult harbor seals, differences in the head shape and facial features can be used to distinguish them. Gray seals have a "Roman nose," with the eyes set well back on the head. In front view, the nostrils appear to form a W. In contrast, the more dog-like face of the harbor seal has the eyes farther forward, and the nostrils form a heart or V shape when seen head-on.

Most fishermen hate gray seals. The seals eat some of the same species of fish that are the targets of fisheries—salmon, herring, cod, and haddock, for example. They also mutilate and remove fish from nets, damaging gear in the process. Finally, the gray seal plays a key role in the life cycle of a parasitic worm that infects cod. Although the worms are not particularly dangerous to humans, fish processors assume that consumers would be outraged if they were not removed. The extra handling time adds to the cost of getting the fish to market.

In response to the claims against gray seals by fishermen, the Canadian government sponsored an annual cull of about 1000 gray seals (mostly first-year pups) from 1967 to 1983. Starting in 1976, Canada also offered a bounty on gray seals, which meant that an additional thousand or so animals were killed each year until 1983. Although no organized kill of gray seals occurs today, they are fair game for Canadian fishermen except during the winter pupping season.

Despite Canada's efforts to keep the population from growing, gray seals have continued to increase their numbers and expand their range. Today, the total population in the western North Atlantic is approximately 200,000. The number in Maine increased from about 30 in the early 1980s to between 500 and 1000

△ This adult gray seal exhibits the characteristic "Roman nose" that helps distinguish gray seals from harbor seals.

△ Gray seals haul out on breeding island.

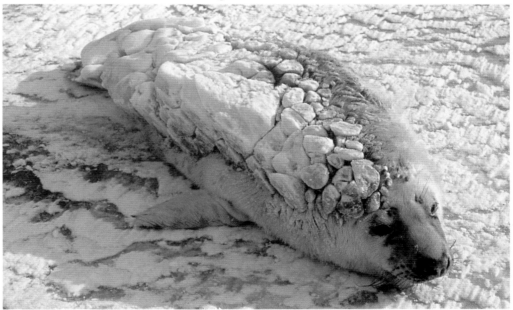

△ This gray seal pup, its hair encrusted with snow, is sufficiently insulated to endure the harsh winter typical of New England and the Canadian Maritimes.

in the mid-1990s.

Of particular interest to Americans is the colony of gray seals that pups on Muskeget Island, west of Nantucket Island off the coast of Massachusetts. Although no more than 13 gray seals had been counted at Muskeget during any one year in the 1970s, more than 1500 were counted there in 1993.

One reason for the gray seal's ability to increase so rapidly is its tolerance for various pupping habitats. Gray seals give birth on everything from rocky mainland shores and sandy island beaches, to the surface of sea ice. Their peak season for pupping in the western North Atlantic is mid-winter. Gray seals, unlike harbor seals, are born in a silky natal coat (lanugo) and do not molt until two to four weeks of age. They are weaned and finished molting by the end of their first month of life, after which they usually fast (living off their blubber reserves) for several weeks.

HARP AND HOODED SEALS
(Phoca groenlandica
and *Cystophora cristata)*

△ Motorists in Tenants Harbor, Maine, expect to see wildlife on their roads, but usually not hooded seals. This one stopped traffic in early February 1995.

These two species are the most controversial seals in the North Atlantic, made famous by media coverage of the bloody commercial harvest of white-coated and "blue"-coated pups (newborn harp and hooded seals, respectively) on the ice off eastern Canada in the 1960s and 1970s. Although it was scaled back for a number of years because of public outrage, the killing of hundreds of thousands of seals resumed in the mid-1990s.

Twenty years ago, when the populations of harp and hooded seals in the western North Atlantic were being kept down by the killing, including them in a book about mammals of the northeastern United States might have seemed questionable. In those days, a solitary harp seal occasionally wandered away from the usual path of migration and found itself on a beach in New England. Slightly more often, late winter visitors to a Maine beach might have discovered a female hooded seal with her newborn pup—far off course.

Today, however, the situation is dramatically different. With a Canadian harp seal population of about three million, this species has become a regular visitor to US waters. During the four years 1989 to 1992, an average of eight harp seals were found each year on beaches in the northeast. This average leaped to 45-50 during 1993-1994. Carcasses

△ This photo of a harp seal shows why the annual slaughter of thousands of pups attracted so much attention and sympathy from the public. By the 1980s, pressure to ban the clubbing of unweaned pups was sufficient to force the Canadian government to comply. Although the commercial hunt for harp seals continues on a large scale, it no longer includes the killing of unweaned pups with clubs.

have been recovered in recent years not only from Maine and Massachusetts, but also from Connecticut, New York, and New Jersey.

A similar pattern has developed with hooded seals, of which there are now close to half a million in the western North Atlantic stock. The average number found stranded on US beaches increased from seven per year in 1989-1992 to 19-24 in 1993-1994.

△ A harp seal raises its head through a hole in the sea ice in Canada's Gulf of St. Lawrence, where many thousands gather to pup and mate each spring.

△ The impressive red membrane on the male hooded seal's head inflates during courtship or when the animal is threatened. In addition to making the male more noticeable and possibly more attractive to females, it greatly increases the apparent size of the head, which may make the seal's enemies think twice about attacking.

THE ICE SEALS

Both harp and hooded seals are "ice seals," meaning that they haul out and give birth on sea ice. They assemble during winter (late February to mid-March) in huge "rookeries" on the stable ice in the Gulf of St. Lawrence and off the Labrador and New-foundland coasts. Pups are born in the attractive pelts that sealers covet. After a brief nursing period—less than two weeks for harp seals and only four days for hooded seals—the pups are on their own. The young harp seals molt on and around the ice floes, then head north to join juveniles and adults on the summer feeding grounds off Green-land and the Canadian Arctic. The young hooded seals, having already shed their fetal coat before birth, go north before molting again. After leaving the North American coast, juvenile and adult hooded seals assemble on the thick ice floes along the south and east coasts of Greenland for molting before proceeding northward to feed inten-sively in summer and fall.

Where to See Mammals in the Northeast

MAINE

Baxter State Park, Millinocket

Sandy Stream Pond and Campground is one of the best locales in the lower 48 states for observing moose at close range. These huge deer are very much accustomed to the visitors hiking and canoeing along the lake shore. When the moose come to feed in the lake shallows in the late afternoon, it's easy to get quite close.

Acadia National Park, Bar Harbor

Ask the rangers about the latest beaver dam activity. Beavers are usually common and fairly tame along some of the feeder streams where they build small dams in the birch groves. Along the coast, watch for harbor seals.

Moosehorn National Wildlife Refuge, Calais

Moose, deer, otters, bobcats, red foxes, and coyotes can be seen, as well as harbor seals along the coastline.

VERMONT

Atherton Meadow Wildlife Area, Whitington

Look for white-tail deer, red foxes, raccoons, star-nosed moles, and muskrats.

MASSACHUSETTS

Whale-Watching

A number of whale-watching tours are conducted from the towns of Gloucester, Plymouth, and Provincetown and from the New England Aquarium in Boston. Humpback and fin whales and dolphins are the usual highlights of these tours.

Monotony National Wildlife Refuge, Chatham

These two small islands host a huge winter population of harbor seals, and a small number of gray seals.

Great Meadows National Wildlife Refuge, Sudbury

White-tail deer, mink, long-tailed weasels, raccoons, and muskrats can frequently be seen.

CONNECTICUT

Rock Spring Wildlife Refuge, Scotland

Beaver, deer, and muskrats.

McLean Game Refuge, Grandby

White-tail deer, gray squirrels, and eastern chipmunks.

NEW YORK

Connetquat River and Montauk Point State Parks, Montauk, Long Island

White-tail deer are everywhere in the Northeast, but they're most easily watched or photographed where they are safe from hunting. Connetquat River State Park has a large herd of very friendly deer, as well as wild turkeys, that are virtually oblivious to the presence of humans.

New York Zoological Gardens, Bronx

The Bronx Zoo has several native northeastern mammals on display. Also, it has a very photogenic population of the black morph of the eastern gray squirrel.

Constitution Island Marsh Sanctuary, Garrison

Skunks, red foxes, raccoons, opossums, mink, and muskrats.

Iroquois National Wildlife Refuge, Alabama.

White-tail deer, red foxes, muskrats, woodchucks, and beavers.

NEW JERSEY

Island Beach State Park, Seaside Park

The wily red fox is found throughout the region, but to avoid conflict with man, it usually stays out of sight during the day. This is not the case here, especially in winter, when red foxes of several color phases can be seen along the park road foraging for edible scraps of trash.

Briantine/Forsythe National Wildlife Refuge, Oceanville.

Best known for its birds, this large refuge contains fresh and salt marshes, fresh water impoundments, scrub oak and pine forests with a variety of elusive mammals. Muskrats are common in the fresh water marshes as are white-tailed deer in the oak and pine forests and surrounding fields. Red squirrels flourish in the pine woods too, and river otters cruise wherever there is fresh or saltwater. On the auto tour, it's not unusual to see a river otter loping along the road.

Green Swamp National Wildlife Refuge, Basking Ridge

White-tail deer, red foxes, raccoons, opossums, muskrats, and weasels.

PENNSYLVANIA

Elk County, Ridgeway

Although elk were once found throughout the Northeast, native herds were extirpated long ago. Elk County has a thriving herd of reintroduced elk or wapiti. They are visible along the roadside at dusk or dawn. Wapiti feed in meadows or along the open areas near the sides of rural roads. The herds move about a bit, so it is best to check with the locals. Ask at the diner or sporting goods store.

Ricketts Glen State Park, Benthon.

This park is best known for its wonderful series of cascading waterfalls. It attracts plenty of summer campers. Watch for the deer when they are fed in the early morning or late afternoon around the park headquarters. During the day, the deer are best seen along the grassy bottomlands below the dam breast. At night, look along the roads and campground for black bears.

Erie National Wildlife Refuge, Guys Woods.

White-tail deer, cottontail rabbits, muskrats, and red foxes are fairly common and approachable on a long strip of park that juts into Lake Erie.

Penn's Wood Exhibit, Philidelphia Zoo, Philadelphia

This exhibit meaders through typical forest habitat. It is a good place to see captive bobcats, black bears, deer, and foxes in a fairly natural-looking setting.

Tinicum/John Heinz National Wildlife Refuge, Philadelphia

White-tail deer, muskrats, meadow voles, and red foxes.

DELAWARE

Prime Hook National Wildlife Refuge, Milton

Red and gray foxes, muskrats, white-tail deer, and an endangered sub-species of fox squirrel, the Delmarva fox squirrel.

MARYLAND

Blackwater National Wildlife Refuge, Campbridge

White-tail deer, nutria, and Delmarva fox squirrels.